Is that Fish in Your Tomato?

The Fact and Fiction of GM Foods

Rebecca Nesbit

Published in 2017 by Ockham Publishing in the United Kingdom

ISBN 978-1-910780-16-9

Cover design by Armend Meha

www.ockham-publishing.com

Acknowledgments

I would never have reached the point where I could write this book without the support of my family, and a particular mention should go to my mother, Judy, who provided early feedback on my draft. I'm also grateful to Siân Deller and Rob Spencer for valuable feedback on the manuscript, along with Rob and Sarah from Ockham Publishing.

I receive support from far more friends than I could mention, but those who particularly encouraged me in writing this book include: Suzannah Watson, Beatrix Li, Charlotte Bratt, Lisi Bouchard and Valli Murthy, and of course my husband Phil Gould.

Many people gave their time freely for interviews and discussions which fed into the text and helped shape my understanding – I am extremely grateful to them all.

Ultimately, I would like to thank everyone whose research has brought evidence to the GMO debate, and everyone who has used this evidence in discussions, even when it is inconvenient. Thank you in particular to everyone who shares the information they read in this book with family and friends.

Contents

Chapter 1

Introduction

Two decades after the first GM foods went on sale, many news reports go something like this:

Reporter: Environmentalists have today released a report which shows the devastating effects of GM crops on farmland wildlife.

Camera zooms in on activist on the couch.

Activist: Yes, this is further proof of the dangers of genetically-modified organisms. The only people who benefit are the companies who developed this Frankenfood, and the rest of us pay the price.

Reporter: And how has the company responsible for this technology responded?

Cut to industry representative in a different studio.

Industry representative: I'm afraid the conclusions of the report are simply incorrect. It completely misrepresents the reality, which is that GM crops pose no more risk to health or the environment than conventional crops. In fact, they bring major benefits.

Cut back to the couch.

Activist: I'm sorry but that's wrong. You are ignoring studies which link GMOs to cancer and to terrible environmental damage.

We are gambling with our future by releasing untested and dangerous new technologies into the environment.

Industry representative: Any GM crop on the market has gone through years of rigorous testing. Major scientific societies around the world have agreed that GM crops are safe. And more than that, we are going to need to use genetic engineering to feed a population of 9 billion people.

Cut back to the news reporter who is thinking 'job done, we've presented both sides of the argument'. You reach for the remote feeling none the wiser.

Life would be easier if we could firmly hold an extreme position on GM crops – it would be reassuring to have a conviction either that GM crops should be stopped at all costs or that they will solve world hunger. At risk of a spoiler alert, the evidence doesn't conveniently support either of those world views. In the following 19 chapters I will look at the complexity surrounding the possibilities, the risks and the limitations of genetically modified organisms. We will cover everything from organ-donor pigs to purple tomatoes, and consider challenges which range from 'superweeds' to patents.

This book stems from my dissatisfaction with the 'facts' I was presented with by people on both sides of the debate. GMOs had always captured my attention, and I began to explore the science whilst working at Rothamsted Research, the world's oldest agricultural research station. I was an ecologist studying butterfly migration, but was exposed to some of the complexities of creating a sustainable food supply. The stakes are high – few things are more important than our health and the environment. The concerns were too great to ignore, though I needed to throw out some long-held convictions.

Previously, the very negative view from environmental campaigners had rung true for me. Around the turn of the millennium, I was doing everything that an environmentally conscious teenager should do: I challenged the school's recycling policy, attended Friends of the Earth meetings and argued vehemently against GM crops. As I learnt more, my opinions became increasingly sophisticated. During my time as a biology undergraduate, I wrote a very convincing essay arguing for a low-risk strategy when dealing with the environment. GM crops, I reasoned, aren't a gamble we should be taking.

However, as the years went on an extreme position became increasingly hard to justify. As I learnt of the risks we're taking with our current food production, I realised that we can't afford to be idealistic about solutions. At the same time, our understanding of the effects of GM crops was increasing. Some people argue that we released the first GM crops without enough evidence that they were safe. Whether or not this is true, the information we gained from doing so has made it increasingly clear that GMOs aren't inherently dangerous. As more evidence has built up, the American Association for the Advancement of Science and many other scientific societies from around the world have concluded that genetically modified crops present very similar risks to those developed through conventional breeding. Both GM and conventional crops do, however, raise some questions about the environmental and social impacts of food production.

Answering these questions requires knowledge about both science and society, and interpreting the evidence around GM foods is a pretty complex task. For a start, how do you collect all the information you need? Smallholders growing GM cotton don't exactly submit information about wildlife on their land to a central database. And even if we have information about what happened when we introduced GM crops, it's impossible to know what would have happened if we hadn't. Crop yields could be increasing due to improved farm management, for example, rather than due to the GM crop. The same is true when you compare current yields of GM and non-GM crops: are farmers who grow GM crops acting differently to those who don't? Still, there are ways to collect some of this information, and estimates have been made about the overall benefit.

Annual GMO reports from the agricultural consultancy PG Economics provide interesting reading. They present estimated economic benefits of US$18.8 billion in 2012 alone. Their calculations for environmental indicators estimate that insect-resistant crops have reduced pesticide spraying by 503 million kilos, which is a global reduction of 8.8%.

Such estimates are extremely interesting but, even if we could completely trust the accuracy of these global figures, they mask a huge variation in the impacts of GM crops. Different crops have different effects, both positive and negative, and the same is true for different environments and management practices. Economic benefits are also unevenly distributed, both between countries and within countries.

Take insect-resistant cotton, which has brought the greatest benefit where pests are at their worst, as we will see in Chapter 5. Before the introduction of insect-resistant crops, the pest problem either meant yields were drastically reduced or that farmers used large quantities of pesticides. The impact of insect-resistant cotton has often been positive, both for farmers and for insect diversity. There are situations, however, where it hasn't been a success.

In the Andhra Pradesh region of India the initial introduction of insect-resistant crops caused yield losses. Insect resistance had only been put into cotton varieties suitable for irrigated land, yet these varieties were still introduced into areas prone to drought. Predictably, this wasn't a recipe for success. These particular failures were nothing to do with genetic modification itself, yet show a major problem for local farmers which global statistics don't reveal. Just because a crop brings benefits in theory, it doesn't mean it always will in practice.

Even within one success story, new innovations often come with winners and losers. Take Syngenta's product of the year 2014, Enogen corn, developed for ease of bioethanol creation. The company heralded it as a win-win-win solution, benefitting ethanol-production factories, farmers and rural communities (it didn't add the fourth, presumably sizeable, win for itself), and no doubt the crop brought some benefits. Still, even if the alternative is fossil fuels, many would argue that a major loser in the story of biofuels from food crops is the environment. Again, this environmental risk isn't caused by genetic modification itself, but is still a problem associated with the crop.

In parts of the world, growers and seed developers have deemed that planting GM crops will be a win for them, sometimes with incentives from governments. The area of farmland planted with GM crops has been steadily increasing for the last two decades. In 1996, 1.7 million hectares were planted with GM crops, and this had risen to 180 million by 2015. In 2015, GM crops were planted in 28 countries on six continents, and roughly three quarters of this was in just three countries: the USA, Brazil and Argentina. About 12% of the world's cropland was planted with GM crops, and the vast majority of this was commodity crops.

In total, 18 million farmers grow GM crops. Eleven countries in South and Central America planted GM crops in 2014, mostly maize, cotton and soybean. Of the six Asian countries which grow GM crops commercially, cotton is the most common crop, with food crops being approved in just three Asian countries: China, Bangladesh and the Philippines. These include maize, eggplant, papaya and tomatoes.

The situation is very different in the EU, where a small amount of insect-resistant maize is grown, almost exclusively in Spain. However, Europe has approved more GM crops for import, and most of these imports end up in animal feed. In particular, the EU relies on imports for the majority of its soybean needs. With 83% of the land in soybean production planted with GM, it's relatively hard to find a GM free option for this important component of animal feed.

Likewise, few African countries have GM crops approved for planting. Crops which have gained approval in South Africa include varieties of maize, cotton and rice, and GM cotton is grown in Burkina Faso and Sudan. Egypt has approved a variety of GM maize, although in 2015 none was actually grown. A further seven countries are conducting field trials. In 2015, 3.5 million hectares of GM crops were grown commercially in Africa.

The difference between continents is the result of both social factors and technology. So far, commercial crops for industrial agriculture have been the focus of development, with the crops which are important to subsistence farmers attracting much less attention. Currently, almost all the GM crops grown worldwide are resistant to herbicides or to pests, or to both. As we will see, there is a much greater diversity of crops in the pipeline, many of them explicitly created for the developing world.

These techniques are blurring the distinction between GM and 'conventional' breeding. Is it time to question exactly what we class as GM, and consider whether extra regulations for these crops are warranted? Certainly anyone who disagrees with GM because it is 'unnatural' might like to take a look at current plant breeding practices (as we do in Chapter 4). Although the 'natural is best' outlook is prevalent in the GMO debate, it is worth questioning its logic.

'All natural' has become great marketing, but it isn't a way to guarantee benefits either from a health or environmental point of view. Fake fur is undeniably a more environmentally sound choice than catching an arctic fox. Likewise, modern pharmaceuticals are a more environmentally friendly (and effective) treatment for fevers and convulsions than rhino horns. From a health perspective too, natural isn't always best. Even familiar foods such as potato can be harmful in their natural raw state. And unless you've invited your enemies for dinner, you no doubt cut off the rhubarb leaves before making a crumble.

Without 'natural vs man-made' as a simple way to judge risks and benefits, we instead have to rely on the evidence. Specifically, we need to look at the evidence for each crop on a case-by-case basis. There's a wide variety of GM crops under development, so we can't conclude that all GMOs are good or all GMOs are bad. The differences between GM crops aren't simply scientific; they are also social and economic. Who owns the technology, for example, and is it being used in a responsible way?

Many of the issues we have to consider aren't unique to GM, even though they are often presented that way. The GM debate, so often fuelled by misinformation, can distract us from the real issues of sustainable food production. These issues are vitally important. Agriculture is by far the leading cause of deforestation, is responsible for 70% of freshwater extraction, and causes about a third of greenhouse gas emissions. Faced with a growing population and a changing climate, we have some very serious challenges to meet. Meanwhile, the current GM stalemate is draining resources both from people developing GMOs and from their opponents.

We have come a long way in the two decades since the first GM crop was commercialised, and thankfully the more dramatic early predictions certainly haven't played out. In March 2000, Greenpeace released a briefing on GM crops entitled 'The End of the World as We Know it', accompanied by the tagline 'Don't do it'. It's not entirely clear what the authors were envisaging, but I'm pretty certain the title was meant to indicate a more apocalyptic scenario than the one we've seen. The world does look decidedly different to the one we saw at the turn of the millennium, but GM has yet to cause a revolution. Instead, as we will see, it has brought both benefits and problems, and has the potential to bring more of both. With potential effects on our food system and our environment, the issue is too big to ignore.

Chapter 2

Birth of the 'Frankenfood' Debate

In 1992, New York Times Magazine's food columnist Molly O'Neill predicted that GM food "could be the biggest boon to corporate profits since frozen foods were introduced in the 1930's, and it could also be a marketing nightmare".

As the end of the 20th century came closer, it became clear that only one of these predictions was proving correct. It wasn't the 'best thing since frozen peas' for the food industry, but this would be the start of a PR struggle lasting decades. Destruction of GM crops became a popular pastime for 1990s environmentalists, with protests as creative as making crop circles in the shape of question marks. Similarly, the biotech industry fought back.

Anti-GM sentiments had started brewing in the 1980s, and in 1991 Friends of the Earth set up a biotechnology programme to "inform policy-makers of the risks to the environment from genetically modified organisms". It was only in 1994, however, that things really got interesting. This year saw the launch of the world's first genetically modified food, and it was a hit.

It wasn't a multi-national giant that was first to put GM food on the shelves, but a small, close-knit company from Davis, California. Calgene had created the Flavr Savr tomato, which looked set to be a poster child for biotechnology and to turn its inventor into an industry leader. Belinda Martineau, who helped bring Flavr Savr to market as a principal scientist at Calgene, has shared the story, warts and all, in her book First Fruit. She describes a journey of technical challenges, complex safety tests, regulatory hurdles and financial calculations that didn't add up. Eventually, Flavr Savr provided the fame but not the fortune.

The tomato plant hasn't made it easy to deliver ripe and tasty delicacies to your dinner plate. As any greenhouse owner knows, pick them too late and they rot, pick them too early and it's green tomato chutney all round. The tomato industry has risen to the challenge by picking green tomatoes and then using ethylene gas to induce ripening. There's little choice in the matter: tomatoes harvested with even a hint of red on them won't survive the shipping. Artificial ripening does, however, come at the expense of flavour.

A tomato which doesn't go off sounds like the holy grail for farmers, retailers and tomato lovers, and perhaps the Flavr Savr could be their dream come true. The altered gene in Flavr Savr coded for an enzyme that breaks down pectin, which is what causes the fruit to soften and rot. In Flavr Savr the gene was flipped, meaning less pectin and a tomato that is much slower to rot.

In the late 1980s Calgene produced its first mature Flavr Savr tomatoes, and tests showed these had the long shelf life the scientists had been striving for. The centrefold in Calgene's 1989 annual report was a photograph of juicy Flavr Savr tomatoes alongside their rotting non-GM counterparts. So far, they were on track for success. It was time to tackle the extensive testing needed to satisfy the regulators.

In the USA, regulation of GM foods is the domain of the Food and Drug Administration (FDA). Although Calgene could be held responsible if their foods were deemed unsafe, they were under no obligation to seek FDA approval before releasing Flavr Savr.

However, they considered FDA approval imperative for public relations so made the commendable decision to make all the data supporting the application freely available to the public, whatever they found.

As the first company to enter the brave new world of commercialising GM foods, Calgene had to deal with the regulation both of general techniques of genetic modification and of the specific modifications made to Flavr Savr. To make sure the journey to market was as smooth as possible, they decided to plough ahead with approval for GM techniques while Flavr Savr was still under development.

The most worrying technique was the use of a 'marker gene' which made the GM plants resistant to the antibiotic kanamycin. These marker genes are used to identify which experimental plants contain modified DNA. By inserting an antibiotic resistance gene along with the desirable gene, scientists could use the antibiotic to detect whether the plant had been successfully modified.

In hindsight, resistance to an antibiotic used to treat humans might not have been a smart choice, especially for public relations. However, by this time Calgene had thousands of prototype plants containing the gene so Belinda and colleagues were committed. She explained: "We did not discuss these issues out loud. Rather, we seemed to silently agree there was no looking back."

The only choice was to convince the FDA that this gene was both necessary and that it was safe to consume.

In 1990 Belinda was one of the scientists who reluctantly accepted the task of doing the experiments needed to satisfy the regulators. She was convinced by the stark realisation that "without the regulatory approval we were seeking and the subsequent successful commercialisation of the Flavr Savr tomato, the stability of the entire company was in jeopardy. What we were really undertaking was science for survival."

A question which clearly needed addressing was whether Flavr Savr tomatoes could make bacteria in the human gut resistant to kanamycin. This task fell to Belinda, and with trepidation she embarked on experiments to determine how thoroughly DNA gets broken down in the digestive system, and whether the kanamycin-resistance gene could jump from the tomatoes to bacteria. She survived the smell of the synthetic gastric fluids used in experiments and was rewarded with the results she'd hoped for.

Even making many worst-case scenario assumptions, she calculated that for every 1,000 Flavr Savr consumers one gut bacterium might become resistant to kanamycin. This doesn't sound too bad given that one person's gut contains around 10^{12} bacteria which could theoretically take up the gene. What's more, other studies were showing that kanamycin resistance was already common in human gut bacteria. Her colleagues came to the same conclusions for bacteria in agricultural soil, which was important for the environmental impact assessment. Spirits were high when the kanamycin assessment was sent to the FDA in 1990.

Meanwhile, the media was taking an interest in the potential for Flavr Savr tomatoes, to the delight of Calgene's business department. In their excitement they decided that promises of longer shelf life weren't enough. Their marketing claims went a step further, to say that the tomatoes would be firm enough to survive shipment even if they were harvested when they were ripe. And to truly place Flavr Savr as a premium product, they claimed that they would taste better than standard tomatoes.

It seemed a logical assumption, but Belinda described the scepticism of some of her scientific colleagues: "It would have been one thing if we were simply claiming that the Flavr Savr tomato could linger on the grocer's or consumer's shelf weeks longer than the typical fresh tomato. But a vine-ripened fresh tomato that could survive the shipping process? That tomato was purely hypothetical."

This scepticism turned out to be wise. The first shipping test was to see whether vine-ripened tomatoes could survive a 2,000 mile truck ride from Mexico. In Belinda's words: "The results of the test were clear before the vehicle had come to a stop. Tomato puree was seeping from the back end of the truck. The cargo was beyond salvage. One Calgene official repeatedly muttered, 'It's over, it's over.' Two others used snow shovels to transfer the mess into dumpsters."

Although Flavr Savr was delivering big improvements to shelf life, the plans for improvements between field and shelf didn't seem to be working out. There was a reason why the industry had stuck to transporting green tomatoes, and Flavr Savr wasn't necessarily going to change this. The team kept working.

Alongside these technical hurdles, the regulatory path was proving to be more challenging than they'd bargained for. The FDA responded to the initial kanamycin assessment with a request for more data. 'Unintended consequences' hadn't been adequately assessed, including the implications of where the new genes were inserted in the tomato's DNA. In many ways GM in the '90s was a pretty crude process. Scientists knew which gene they were adding, but couldn't control where it would end up (the fact that the FDA approval, when it eventually came through, would arrive via fax also gives an insight into the technology they had available in the lab compared to today's biologists). If, for example, the genes were inserted in the middle of the gene for a vitamin, this could change the tomato's nutritional value.

Samples of ripe Flavr Savr were sent to the National Food Laboratory in California, and their vitamin and mineral content was compared to non-GM tomato varieties. Thankfully, analysis showed there was no reason to be concerned by the levels of vitamins or minerals.

To be as conservative as possible in their safety assessment, they also performed an 'acute oral toxicity test'. This is a standard approach to uncover toxic effects on animals. Belinda explained: "Although tomato fruit was a far cry from being a single, pure chemical substance like the ones usually tested using this approach, no one at Calgene could think of a better way to reveal unintended changes we could not imagine."

At the IIT Research Institute in Chicago, pureed tomatoes were fed to healthy rats using a syringe. After two weeks the rats were euthanised and autopsies were performed. The researchers concluded that there had been no toxic effects from the Flavr Savr. These results were promising but preliminary, so were followed by 28 day feeding experiments ('wholesomeness studies'). This time the results looked worrying. Some of the rats fed on one particular variety of Flavr Savr had developed lesions in their stomach linings.

A 'flabbergasted' Belinda was instructed to repeat the tests. All the available fruit from the offending variety was immediately harvested, and this time the fruit samples were freeze-dried to increase their concentration. This way the rats got more fruit, so if the tomatoes were the culprits the lesions would be more likely to occur. The results of this experiment caused Calgene employees to relax a little. Lesions were again found in rats fed Flavr Savr tomatoes, but they were also found in rats fed non-GM tomatoes. What's more, they were found most often in rats fed water rather than tomato. Increasing the concentration of tomato had no effect, further supporting the conclusion that Flavr Savr was not the cause.

To tackle the problem from another direction, Belinda's team decided to investigate whether the lesions could be caused by an increased level of toxins. Tomatoes naturally produce toxins, but these aren't present at high enough concentrations to be harmful to humans (a clear example of the toxicologists' mantra 'the dose makes the poison'). Calgene therefore sent samples to the University of Maine to test for raised levels of tomatine toxin. To Calgene's relief, the results came back negative.

Was the 'good' feeding study enough to negate the previous 'bad' results? It was a question that Calgene employees didn't know how to answer, though as Belinda explained: "One thing was certain. The prospect of carrying out more wholesomeness studies was unappealing. The group's plan of action, therefore, didn't call for any." Pathologists from outside Calgene would analyse information from all the studies, and then it would be up to the FDA to decide.

In this instance, the FDA did accept the evidence as safe, and Calgene's journey into the uncharted territory of approvals for GM foods proved worthwhile. On 18th May 1994 Calgene employees arrived at work to find a banner above the front door reading 'FDA Approval' and at noon champagne was served while a news crew recorded the celebration. This came not a moment too soon for Calgene, which was by now millions of dollars in debt. They lost no time in letting the sales commence.

Three days later, the world's first genetically modified tomatoes arrived on the shelves of just two American grocery stores. Bright-red booklets in the shape of tomatoes accompanied each purchase, briefly explaining the genetic engineering process.

Belinda wasn't the only person to be excited on the launch day, and she described the Flavr Savr's first appearance: "[At] the store in Davis, where Calgene is located, they had to ration the tomatoes – the store owner said you could only buy two a day. And literally, Calgene couldn't keep up with the demand. They were that popular."

This enthusiasm continued, and the following May Flavr Savr tomatoes were shipped to 1,700 stores, selling for up to twice as much as other tomatoes. Their popularity was fuelled partly by the success of a blind taste test in which employees of high-end restaurants declared their surprise at the sweet and juicy tomatoes.

Sales may have been looking good, but Calgene continued to report losses of millions of dollars. Flavr Savr, meant to help them back into the red, wasn't helping. They still hadn't properly solved the problem of the tomato's ability to survive shipping once ripe, and so transportation was difficult and expensive. This was compounded by the challenges of harvesting ripe tomatoes: you can pick green tomatoes together, but ripe tomatoes are ready for harvest at different times. Once you took shipping costs into account, Flavr Savr was actually selling at a considerable loss. Calgene's lack of experience in the tomato business was painfully obvious.

Its financial plight prompted Calgene to accept a first investment from Monsanto in 1995, although that wasn't enough to reverse the trend. Finally, in 1997, Monsanto bought all remaining shares. This marked the end of the road for the Flavr Savr.

Despite this disappointment for everyone who had worked to make the Flavr Savr dream a reality, their experiences of regulation and commercialisation weren't in vain. They had paved the way for others in the industry, and the mid-1990s were exciting times for biotech. Other companies were hot on Calgene's heels in bringing GM foods to market. While Calgene's plans for fresh tomatoes began to unravel, the story with tomato paste was panning out quite differently.

The first GM product to go on sale across the Atlantic in the UK was a tomato paste. It was made with tomatoes developed by Zeneca which, like Flavr Savr, had a longer shelf life. This modification meant the paste could be produced more cheaply, no doubt one of the reasons it proved popular. Despite the front of the tins announcing in large letters that the paste was made with genetically modified tomatoes, over 1.8 million cans were sold in British supermarkets between 1996 and 1999.

During this time, however, the mood began to change. Initially, the biotech industry was largely unconcerned by public opinion. Throughout the 1980s they'd been warned about the dangers of public resistance both to milk from hormone-fed cows and to GM. When the 90s came along and none of this had fully materialised, it looked a bit like crying wolf. And for a while, things progressed very well indeed: in 1996 Monsanto shareholders got a 62% return on investment.

There were worrying signs, however. The protesters who paid Calgene a visit ranged from Greenpeace to the Union for Concerned Scientists. Predictably, some of the opposition was accompanied by misinformation. One rumour claimed that the tomatoes were cubes (now that would be a shipping company's dream), another that a fish gene had been added, with potentially dire consequences for anyone allergic to fish. The latter rumour provided great material for anyone wishing to illustrate genetic modification as 'playing God', and some imaginative images of fish-tomato hybrids sprang up.

Still, it wasn't the protestors who had put the nails in Flavr Savr's coffin. Even though the tomatoes were labelled as GM and were more expensive than other premium tomatoes, consumers were not deterred. Most environmental advocacy groups agreed that Calgene had done a good job of demonstrating Flavr Savr's safety with publically-available data. Belinda explained: "There were many reasons why the Flavr Savr tomato eventually flopped but public outcry at the fact that it was genetically engineered was not one of them. Almost without exception during the course of its brief commercial run, demand for the Flavr Savr tomato outdistanced supplies."

This trend was not to continue. Despite the initial positive sales, the level of public resistance in Europe became hard to ignore. Not a problem, thought those at the top, the public just needs to be educated. In 1996, Monsanto opened information hotlines in the UK and Germany, with little effect. As the softer attempts at education failed, Monsanto embarked on an advertising campaign which would be a monumental disaster. Looking at the ads, it's not hard to see why.

The headline of one advert managed to simultaneously belittle those who cared about world hunger and over-hype the role of biotech: "Worrying about starving future generations won't feed them. Food biotechnology will." To counter images of mutant vegetables, Monsanto tried to play the 'natural' card. The advert explained that GM seeds had "naturally occurring beneficial genes inserted into their genetic structure". I was half expecting them to go on to give one of these genes a positive name, in the same way that hair care adverts invent names for molecules in shampoo.

The hype around new technologies regularly leads to disillusionment – I'm yet to own a 3D printer, and it turns out that Apple's intelligent personal assistant, Siri, does not in fact know everything you need. Early promises are often not fulfilled, or at least not quickly enough, and in this way Monsanto helped set the biotech industry up for a fall. One advertisement promised: "Less chemical use in farming, saving scarce resources. More productive yields. Disease-resistant crops."

In addition to pro-biotech propaganda, they did try to address some of the dissenters' concerns: "Of course, we are primarily a business. We aim to make profits, acknowledging that there are other views of biotechnology than ours." I have no idea what the acknowledgement meant in practice, or how this was meant to allay any fears. Needless to say, it didn't.

As well as adverts, the US$1.6 million European media blitz included leaflets, hotlines and a consumer website. It claimed to be providing consumers with "the information they need to make informed decisions".

Plenty of people had already made up their minds and didn't feel they needed Monsanto's input into their decisions. Two days after the launch of Monsanto's campaign, the UK's Prince Charles wrote an essay in the Daily Telegraph making associations between BSE and GM in terms of unpredictable consequences, unknown effects and uncertain science. To him, GM crops were "enough to send a cold chill down the spine". This debate was not going away.

In 1998 rifts between the two camps grew deeper following the claims made by Dr Árpád Pusztai. Hungarian-born Pusztai had spent time in a refugee camp in Austria before dedicating 36 years to studying plant proteins at the Rowett Research Institute in Scotland. He started a media storm with an interview on the UK's World in Action TV show where he told viewers he wouldn't eat GM food. It's no surprise that he caught the world's attention by saying "I find it very, very unfair to use our fellow citizens as guinea pigs."

Pusztai had done a study feeding rats with potatoes modified to produce a pesticide which occurs naturally in snowdrops. He told the media that the rats fed GM potatoes were less healthy than those fed non-GM potatoes. Many scientists loudly disagreed, and so the 'poisoned rat debate' began.

In his experiment, the organs of the rats that had eaten GM potatoes weighed less than the organs of those fed non-GM potatoes, and their lymphocytes (immune cells) were depressed. Pusztai showed that rats could safely eat the pesticide on its own; in fact it even protected them against salmonella. He therefore concluded that it must be the new gene itself, or DNA inserted with it, that was the problem. His message was that "the damage to the rats did not come from the lectin, but apparently from the same process of genetic engineering that is used to create the GM foods everyone was already eating".

The media interest in the days following the World in Action broadcast was intense, and the story was covered on TV and radio stations around the world. Pusztai started to became very uncomfortable about what he was hearing on the news. "I heard things that really disturbed me," he said. "My head was buzzing ... the whole thing was getting totally out of hand."

Rebecca Nesbit

Meanwhile, scientists at the Rowett Institute reviewed Pusztai's data and discovered that some of the experiments he'd spoken of weren't even finished, to the alarm of the Institute Director, Professor Philip James.

Following the discovery that the experiments hadn't actually been completed, Pusztai was suspended and misconduct procedures were used to seize his data and ban him from speaking publicly. His annual research contract was not renewed, although he was offered the chance to continue as a lecturer. At the age of 69, we would retire with a cloud over his career.

Preventing Pusztai from speaking publically did nothing to halt the flow of media coverage. The debate took a turn for the worse with claims the potatoes were modified with a gene from jack bean that is poisonous to mammals. This misinformation was formalised in a press release issued by the Rowett Research Institute. James says this press release was approved by Pusztai, though Pusztai says he didn't see it before it went out.

Whatever the status of the press release, Pusztai had been given permission to do the initial interview, and their press officer was present at the start of filming. Looking back on it a decade later, Dr Pusztai described James as excited by the media attention: "The director kept running around like a blue-arsed fly. This was a tremendous public relations business for him."

Exactly what happened in the early stages of the debate is unclear; James claims to have had grave doubts about the interview, but afterwards he called Pusztai to congratulate him. Pusztai has described James's subsequent behaviour as irresponsible: "Apparently he thought the best way to extricate himself from the responsibility for having misled the public … was to tell the world that I got 'muddled' or even that I 'took' data from a colleague who was absent at the time."

21

Pusztai was widely condemned for going public with his conclusions before they were published in a scientific journal. The GM potato was an experimental model and wasn't available as food, so their urgent publicity wasn't needed to protect consumers. When giving evidence before the UK government's Science and Technology Select Committee in 1999, Pusztai estimated that his experiments were 99% complete. What was missing, however, was the chance for other scientists to take a critical look at his work.

One of the highest profile criticisms of the study came from the Royal Society in London. A committee of experts concluded that Pusztai's experiments were badly designed and the statistics he'd used were inappropriate. One major flaw was that a diet of raw potatoes is bad for rats regardless of whether or not the potatoes are genetically modified. Another possibility is that that the effects he was seeing were due to changes in the potatoes occurring because of a different laboratory process used, tissue culture, and not the introduced gene at all. As a result, many scientists see the Pusztai affair as disproportionately damaging to the GM debate.

Professor Chris Leaver, a plant scientist at Oxford University, believes that NGOs decided to make a play using him. He said: "I think it did a lot of damage because ... the vast majority of people were somewhat neutral at the time. I think he got hijacked and then he got out of his depth."

When giving evidence to the House of Commons Select Committee on Science and Technology, the Rowett Institute summarised the way many scientists felt: "Dr Pusztai's concerns about the need for devising new safety tests for transgenic lectins are, in our view, valid. We judge, however, that his experiments to date are far too crude and preliminary to justify any claims for novel findings of either lectin-related or general biotechnological significance."

Amongst the criticisms, Pusztai was even rumoured to live in a gothic mansion bought with the proceeds of guest appearances as an eco-champion. The mansion is one of the easiest fantasies to dismiss in the story of the GM debate, though he has since given hundreds of lectures around the world.

To many anti-GM campaigners, Árpád Pusztai is a hero. He gave the public information that they needed to hear even at the expense of his career. As Pusztai himself said: "I was publicly funded and I thought the public had a right to know."

In this messy debate it is easy to forget to disentangle the different arguments. There are lots of scientific reasons to doubt Pusztai's conclusions, and I will discuss the safety of GM foods in Chapter 10. But even if we don't accept his claims, it doesn't mean he was treated fairly. And likewise, even if he wasn't treated fairly it doesn't mean that the whole scientific community has a policy of crushing anyone who expresses doubts over GM. The way the fiasco was handled was arguably heavy-handed, but we need to look at the safety of GM foods and the way the scientific community communicates its findings as separate issues.

It is also interesting that Pusztai gets all the blame for the debate being more damaging to the image of GM than the data warranted. Speaking as a former science press officer, I would be mortified if this had happened on my watch. And as for those who disciplined Pusztai, could they have done more to put forward an accurate message?

The change in the public mood in the reaction to Pusztai's potato-eating rats was reflected in the BBC's 1998 background briefing on genetically modified food. This focussed on GM's potential as an 'unseen threat'. Naming supermarkets which stocked the tomato paste, the BBC warned that "scientists are ready to bombard the world with genetically modified food". Such was the level of concern that the European Union placed a moratorium on the production of GM crops from 1998 to 2003. It was still legal to import certain approved GM products, but by the second half of 1999 all of the major supermarket chains in the UK had withdrawn GM foods.

The strength of the anti-GM backlash and the level of hype surrounding Pusztai's claims prompted the UK Parliament's Science and Technology Committee to produce its first report on GM foods. The report, published in 1999, highlighted their fear that: "GM technology and its potential benefits may be permanently lost to the UK unless there is rational debate."

The report is far more positive than the mood at the time, and this wasn't welcomed by many campaign groups and members of the public. It did, however, outline the potential benefits and risks, and recognise that scientific findings don't always produce the clear-cut answers the public, press and policy-makers are looking for. Some of their general conclusions are still worth remembering today:

"No human activity is entirely risk-free. Certainly no food is. This is not, however, a reason to trust unquestioningly in new technologies such as modern genetic modification. Risks must be identified, evaluated and minimised.

"It would be deeply regrettable if the UK forfeited all the potential economic and social benefits offered by GM technology on the basis of unfounded scare stories. If the UK is to reject it, it should be on the basis of scientific assessments of identifiable risks or well-considered value judgements not the result of journalistic hyperbole and unfounded fear."

The Committee's worry about scare stories was certainly playing out. As the 20th century drew to a close, the hypothetical concerns of those who spoke out against GM were gradually being backed up with more case studies. Unfortunately, the debate has been shaped partly by stories that turned out to be untrue.

In 1999, it was the turn of the monarch butterflies to take the centre of the GM stage. The monarch, a beautiful butterfly common in the Americas, has an incredible life cycle. Caterpillars which eat their way to adulthood in the USA will migrate south in the autumn as butterflies. They then spend the winter in Mexico's cloud forests, with tens of thousands clustered together on individual trees. Come the spring they will begin a northwards journey of thousands of kilometres and then lay their eggs so the cycle can continue. There are fears, however, that this may be under threat, with monarch populations declining at an alarming rate.

Monarchs have captured imaginations around the world – they have even been taken to the international space station – so it's not surprising that a study suggesting they were being killed by GM crops hit the headlines. The USA was widely growing Bt maize, which has a gene inserted for a naturally-occurring insecticide. The study claimed that pollen from Bt maize was poisonous to monarch caterpillars. Although monarchs don't eat maize, the pollen gets blown onto the caterpillars' foodplant, milkweed.

Claims that butterflies were being poisoned by pollen didn't hold up to scrutiny because the laboratory study used far higher concentrations of pollen than monarchs encountered in the wild. However, by the time this was established the monarch had become a flagship for the dangers of GM crops. Agriculture is indeed a major factor in monarch declines, particularly with reductions of milkweed on farmland. However, there is no simple link between GM maize and the monarch's troubles.

What's still open for debate is whether a more meaningful version of these experiments should have been carried out before Bt maize was grown commercially. However, even if such early experiments had been undertaken to show that toxicity of Bt maize wouldn't threaten the monarchs, this doesn't solve controversies of indirect effects. Could GM crops cause an even greater loss of wildlife habitat than non-GM varieties? I will return to this story in Chapter 6.

Another pervasive argument against GM which arose in the 1990s is the so-called 'terminator gene'. This controversial technology is designed to support existing plant breeder's rights and patent rights. These 'Genetic Use Restriction Technologies' (GURTs) could allow scientists to create crops which produce sterile seeds. Canadian NGO Rural Advancement Foundation International correctly identified a better PR opportunity with a catchier title, and coined the term 'terminator'.

Most companies view seeds they have developed in the same way the software industry sees its proprietary products. If you want to keep using them, you have to keep paying. This is also true for seeds produced through other breeding methods, and large companies are notorious for suing farmers who save and replant seed without paying royalties. Some form of protection is essential for companies to survive; the money from selling patented seeds doesn't just line the investors' pockets, it is used for future research and development.

Monsanto's website explains its reasons for suing farmers who illegally replant its seeds: "When farmers purchase a patented seed variety, they sign an agreement that they will not save and replant seeds produced from the seed they buy from us. They understand the basic simplicity of the agreement, which is that a business must be paid for its product. The vast majority of farmers understand and appreciate our research and are willing to pay for our inventions and the value they provide. They don't think it's fair that some farmers don't pay."

The problem is that saving seeds is a common practice in many developing countries, where it isn't always practical for companies to collect their royalties. Without this 'brown-bagging' of seeds, some farmers simply couldn't make a living. Biotechnology had taken this seed-saving debate up a notch: if companies create a crop which produces sterile seeds then even the poorest farmers would be committed to paying each year.

The first patent application related to terminator technologies goes back to 1991 and was filed by DuPont. This was granted in 1994, and a terminator technology patent was granted to Zeneca shortly after. The greater outcry, however, came when US taxpayers' money was used for terminator research.

In 1998, the United States Department of Agriculture and the Delta & Pine Land Company were granted a patent for 'seed-embedded protection technology'. A British researcher, Melvin Oliver, had been employed by the Department of Agriculture to lead the project. He justified his work by saying: "My main interest is the protection of American technology. Our mission is to protect US agriculture, and to make us competitive in the face of foreign competition. Without this, there is no way of protecting the technology."

To many, however, this wasn't justified protection of businesses, it was an unethical way to treat farmers. Fierce protests raged worldwide, and the protestors won the battle. In June 1999, as a result of the huge opposition from charities, farmers and the wider public, Zeneca announced that it would not market terminator seeds. A few months later Monsanto followed suit. A de facto global moratorium was soon placed on this technology, which is still upheld today.

This could be seen as a battle won, though to me it leaves a large question unanswered – can we not think of a better way to help the poorest farmers make a living than ensuring they are still able to break the law? There was an outcry at the idea of a technology preventing farmers from saving seeds, yet we already allow this to be done with contracts. Even with terminator genes currently out of the picture, the balance of power between multi-national companies and the farmers who buy from them is a major debate, as we will discuss later on. Commercial crop varieties produced by conventional breeding methods come with restrictions, just as GM seeds do.

It's still worth noting that there is also a potential flipside of terminator genes; they could theoretically be used in the containment of GMOs. If a GM plant can't reproduce, you reduce the risk of these plants 'escaping' into the wild.

Arguments about terminator genes, monarchs and poisoned rats provided fuel for the fire of public disapproval. As a result, the anti-biotech lobby were slowing the development of GM foods. But they weren't preventing it. We entered the 21st century with fireworks, hangovers and a fundamental void between many industry scientists and the consumers they were striving to please.

The scale of this divide is perhaps best illustrated by stories of a Calgene reunion in 2000. Belinda showed photos of her two children, listened to accounts of her friends' new jobs, then broached the subject of the public debate over genetically modified food. She was greeted with enthusiastic renditions of the same arguments which had characterised the industry's attitudes throughout the '90s, starting with 'the public doesn't understand the technology'. Almost 100 million acres had been planted with GM crops in 1999, which her colleagues took as proof that the anti-biotech food crusade had proved ineffective and the controversy would soon blow over.

This certainly wasn't Belinda's assessment of the situation: "I was taken aback. The FDA had received some 35,000 comments from the public in response to its meetings on the regulation of genetically engineered foods held the previous November and December. After sending the FDA only a couple of dozen comments regarding the approval process for the Flavr Savr tomato, the general public in the United States appeared to be making up for lost time. After a moment of stunned silence I therefore informed my scientific acquaintances that they were in denial."

The events of the following decade certainly proved she was right.

Chapter 3

The 21st Century

The stroke of midnight on 1st January 2000 was a victory for technology. The much-feared millennium bug failed to strike, and we were left with working computer systems and computing power growing at an exponential rate. In the agricultural world bugs were of course continuing to thrive, and GM technology wasn't getting the same lucky break. Europe was, and still is, a hub of anti-biotech feelings, and campaign groups actively exported these views to other parts of the world. Perhaps the most shocking GM stories from the early 2000s came from Africa.

In 2002 southern Africa was hit by famine. Flooding followed by drought caused harvests to fail. Combined with deteriorating economic conditions and the high rates of AIDS infection, this led to devastating hunger. The UN's World Food Programme (WFP) responded to the situation with emergency food aid, much of which came from the USA and other donor countries who grow large quantities of GM grains. However, health and environmental fears led to resistance from many of the recipients. Some countries, including Lesotho and Mozambique, took a simple solution to preventing environmental damage: milling the GM grain so that no seeds could be grown.

Others took a firmer stance. Zimbabwe stopped the GM food aid from entering the country, and the Zambian government prevented the distribution of GM grain which had already arrived. With almost 3 million of his citizens facing famine, Zambia's

President Mwanawasa justified the rejection by saying: "I will not allow Zambians to be turned into guinea pigs no matter the levels of hunger in the country."

To try and alleviate fears, a delegation of Zambian scientists and economists were invited on a tour of labs in Europe, America and South Africa. Their report, however, simply caused Mwanawasa to strengthen his views, voicing the remarkable opinion about the GM grains that he would "rather die than eat poison". Others pointed out that GM foods imported from South Africa were available in Zambian supermarkets, and that Americans are suffering no side effects from eating GMOs. Zimbabwe relented and accepted milled grain, while Zambia resisted until it had experienced two more years of famine.

This story revealed yet another way that GMOs were going to affect global food security. It also raised some interesting ethical challenges which go well beyond the GM debate. When governments make life and death decisions for their population, it makes questions of 'who decides' even more pertinent. Should developed countries trust African leaders to make their own decisions, or should we try to influence them? The GM debate has a history of 'interference' by people from developed nations seeking to influencing public opinion. Anti-GM groups will say this is all in the name of education, though a report by the European Academies Science Advisory Council concludes that exaggeration of risks by European sceptics has created difficulties for African policy-makers. Whether or not you agree with the subject matter of this education, it stands to reason that the flow of knowledge to the developing world should be appropriate. Should governments give each citizen autonomy in what they choose to eat? The answer to this question may be different depending on whether we are considering health or the environment. Health risks are borne by the individual consumer, whereas environmental risk is borne by everyone.

Meanwhile, Europe was in the comfortable position of rejecting GM technology whilst still enjoying plenty to eat. The 1990s had seen widespread trials of genetically modified plants in Europe, with a peak of over 250 field trials taking place in 1997. This number fell dramatically in the new millennium, and in 2002 just over 50 took place. The UK stopped trials altogether in 2005.

Commercial approval and planting was also slow, with few European countries growing any GM crops at all. Romania had been Europe's early adopter of GM, starting to grow herbicide-resistant soybean in 1998. In 2006 Romania grew GM soybean over a wide area, but this very soon changed. Romania joined the EU in 2007 and the GM soybeans were not approved for cultivation.

Although these varieties aren't approved for cultivation in the EU, some herbicide-resistant soybeans have been approved for import as food and feed. Soybean is a large component of animal feed and the EU uses far more than it grows. Europe imports large quantities of soy from the Americas, while many Romanian farmers rely on EU subsidies to make a living.

The environmental impact of transporting tonnes of feed half way around the world is an issue, though arguably neither the transport nor genetic modification is the main problem associated with soybean production. The World Wildlife Fund (WWF) puts agriculture, particularly soya and beef, as the biggest threat to the Amazon rainforest. This issue is of far greater concern to Professor Nigel Halford, a plant geneticist from Rothamsted Research in the UK: "You can still get non-GM soybean, but ironically you're cutting down the Amazon to source it. South of the Amazon farmers have gone over to GM, so non-GM soy is planted in the North where the forest is being cleared."

For WWF, it isn't whether soy is GM which determines its sustainability. What's important is how producers protect wildlife, use pesticides and respects workers' rights. Anyone interested in the environmental impacts of soy may be better served by using responsibly sourced soybean such as that produced under the Round Table on Responsible Soy.

European retailers have been keen to source GM-free soybean so they can avoid labelling a product as containing a GMO. Animal feed often does have GM ingredients, however, and meat from livestock fed GM feed doesn't have to be labelled. With 80% of the world's soya now genetically modified, feed that is guaranteed GM free is increasingly rare. It's therefore no surprise that, since 2010, British supermarkets have been quietly relaxing their requirement for non-GM animal feed. The bans they had declared on GMOs in feed have now been dropped, except for organic produce.

It seems somewhat disingenuous that supermarkets had ever introduced these bans. If health was a true concern for supermarkets they wouldn't sell cigarettes, if the environment was a concern they wouldn't have open-fronted refrigerators, and if corporate control was a concern there would be no scandals about how little they pay farmers for milk. However, like publicity around avoiding GM, all these things are good for business.

Although EU consumers and policy makers remained largely resistant, scientific bodies have been very vocal about the potential for GM in Europe. The European Academies Science Advisory Council (EASAC) is formed by the national science academies of the EU Member States and has the remit of providing advice to European policy-makers. Its 2013 Planting the

Future Report stated that "so long as EU policies on farming and the environment are out of alignment with the need to innovate, ambitions to improve agriculture will be thwarted. Its lack of enthusiasm for GM crop improvement has increased the EU's dependency on food and feed imports, and has implications for its scientific research and future industrial competitiveness."

Fears about adaptation to climate change led the authors to argue that improved crop varieties can be created swiftly and more reliably using GM methods. Like many scientific bodies, it concluded that GM crops have no greater adverse impact than any other technology used in plant breeding.

Some politicians were swayed. The British MP most keen to bring the GM debate back into public consciousness was Environment Minister Owen Paterson, who was already unpopular with environmentalists for his handling of the badger cull and his apparent lack of concern about climate change. In June 2012, he made a public speech calling for a more informed discussion about the potential of genetically modified crops. With less than 0.1% of global GM cultivation occurring in the EU, he feels we are missing an opportunity.

I was in the full auditorium listening to his speech, and there were some aspects which were hard to disagree with. It's true that we need both to increase food production and to protect wildlife, and it's also true that we can't be complacent about future food production.

For me, however, it raised questions about what role a politician should have in the debate. He is keen to 'reassure' the public, but will this involve properly addressing people's concerns? Listening to his speech, that wasn't entirely clear: "While I believe that there are significant economic, environmental and international development benefits to GM, I am conscious of the views of those who have concerns and who need reassurance on this matter. I recognise that we – government, industry, the scientific community and others – owe a duty to the British public to reassure them that GM is a safe, proven and beneficial innovation. We must lead this discussion, explaining to the public not only what GM technology is but also how it can help."

His speech was slick, but the way he was whisked off without answering the questions, along with printouts of a word-perfect speech that were waiting as we left the auditorium, made me question how much he actually knew about the topic. His views were certainly less nuanced when not delivering prepared speeches, and decidedly less constructive. He has described opposition to GM technology as 'complete nonsense', and branded golden rice opponents as 'wicked' (an issue we'll discuss in Chapter 9).

It wasn't just politicians who were questioning the European Commission's cautious view – there have been some vocal changes of side. The green movement was forced to take a long, hard look at itself in 2010 when the UK's Channel 4 aired a documentary 'What the Green Movement got Wrong' amidst wide controversy. A panel of former environmental activists reassessed some of the eco-views they had previously held as doctrine, and one of these panellists was Mark Lynas.

Mark was already known for his work on climate change, and in this programme he voiced his support for GM crops. He'd been part of the early debate – he'd co-founded Corporate Watch magazine which was an early publisher on the evils of GMOs and Monsanto, he'd taken machetes to fields of GM crops in midnight raids, and he'd organised the busses for a 1998 sit-in of Monsanto's High Wycombe offices. This all made his latest statements in favour of GM far more news-worthy.

Another major hit of publicity came in 2013 when he gave an apology speech at the Oxford Farming Conference. He started: "For the record, here and upfront, I apologise for having spent several years ripping up GM crops. I am also sorry that I helped to start the anti-GM movement back in the mid-1990s, and that I thereby assisted in demonising an important technological option which can be used to benefit the environment."

He spoke of the increasing disconnect he'd experienced between his science-based analysis of climate change and his willingness to argue against GM crops without ever having read a scientific paper. His exploration of the science had caused his views to do a U-turn and, having learnt from his mistakes, he called upon delegates to "go beyond the self-referential reports of campaigning NGOs".

He finished by saying: "I don't know about you, but I've had enough. So my conclusion here today is very clear: the GM debate is over. It is finished. We no longer need to discuss whether or not it is safe – over a decade and a half with three trillion GM meals eaten there has never been a single substantiated case of harm. You are more likely to get hit by an asteroid than to get hurt by GM food. More to the point, people have died from choosing organic, but no-one has died from eating GM."

Some people may have questioned how influential he really was in the early days of the debate, but reactions to his speech show how influential he is today. A spike in traffic crashed his website and people from around the world joined the debate on Twitter. However, he certainly didn't bring the whole environmental movement with him. Sceptics vocally question his credibility, and they have even accused him of being paid by Monsanto. His Channel 4 GM support ended his ten year friendship with the person who'd been best man at his wedding.

While MPs and campaigners are trying to influence the elec-torate, parts of the electorate are trying to influence politicians. In July 2014, nine environmental charities, including Greenpeace European Unit, wrote to Jean-Claude Juncker calling on him to abolish the position of Chief Scientific Advisor to the President of the European Commission. Juncker, former president of Luxem-burg and incoming President of the European Commission, had already spoken about his intention to review the rules on GMO authorisations. He was new to his position as President and so this struck the charities involved as a good chance to intervene.

The letter stated that: "To the media, the current Chief Scien-tific Advisor presented one-sided, partial opinions in the debate on the use of genetically modified organisms in agriculture, repeatedly claiming that there was a scientific consensus about their safety whereas this claim is contradicted by an international statement of scientists (currently 297 signatories) …"

It is certainly true that Professor Dame Anne Glover, the Chief Scientific Advisor, had previously made her position on GM completely clear, saying: "No scientist will ever say something is 100% safe, but I am 99.99% certain from the scientific evidence that there are no health issues with food produced from GM crops. Just about every scientist I know supports this view. Opposition to GM, and the benefits it can bring, is a form of madness I don't understand."

Calling GM opposition madness doesn't seem any more con-structive than some of Owen Patterson's quips, but it's no doubt true that the vast majority of scientists she meets don't support claims of health risks from GM foods, as we will see in Chapter 11. It strikes me that 297 signatories is embarrassingly few (a Global Warming Petition Project has over 30,000 signatures from 'scientists' who believe there's no convincing evidence for man-made climate change). However, that's a minor point compared to the problem with sacking scientific advisors who have inconvenient opinions.

Despite protest from the scientific community, Professor Glover stepped down and wasn't replaced.

Other parts of the world responded to the evidence and publicity surrounding GM much more positively. By 2013, GM crops were grown in 27 countries in an area totalling approximately 175.2 million hectares. The crops grown are extremely varied, from a carnation modified to make it blue to sweet peppers resistant to a plant virus. The vast majority are resistant to pests or herbicides, the subject of Chapters 5 and 6.

While the public debate rages in Europe, there is little movement in the commercialisation of GMOs. In the EU, approval of GM varieties requires a thorough risk assessment by national evaluation agencies and the European Food Safety Authority (EFSA), "in order to ensure safety for human and animal health and for the environment". Only one variety is currently being grown in the EU, a pest-resistant maize sold by Monsanto. It is grown in five countries and is only a very small proportion of the maize grown in Europe.

There are now, however, a few field trials taking place. In 2012, one of these became headline news when protestors threatened to destroy a trial at Rothamsted Research in Hertfordshire. With the aim of 'decontaminating' the crop, activists formed a new protest group 'Take the Flour Back' (a nod to Rage Against the Machine's song Take the Power Back).

With the threat of vandalism looming strong, four scientists working on the trial put together a YouTube video appeal. One of these scientists was Dr Gia Aradottir, who called upon protestors to please reconsider before "several years' worth of work to which we have been devoting our lives will be destroyed forever".

When the appeal hit the national news, Gia had no idea of the way that it would take over her life. She barely saw the inside of the lab for four months, and found herself talking to everyone from TV crews to allotment societies. Some of these interactions were more positive than others. "I spoke to lots of people who were willing to engage in debates, and even if we didn't always end up agreeing we at least heard each other out," she said. "Take the Flour Back, on the other hand, weren't willing to listen to other arguments."

Take the Flour Back's discomfort with the crop came partly from the fact that it contained a synthetic DNA sequence 'most similar to that of the cow'. It made for catchy marketing – the group's logo was a loaf of bread with a cow's head and legs.

However, Gia found that many of the statements put forward by Take the Flour Back weren't based on knowledge. She also didn't share their concern that the trial was a threat to agriculture. She explained: "There was never any doubt in my mind that this trial was safe. The crop was planted in a tiny area, and we had barriers in place to prevent contamination of other crops."

The crop in question was a variety of wheat which produced an aphid alarm pheromone. The pheromone is not only produced by aphids but also by other plant species, and the gene used in this experiment came from peppermint. Gia sees the modification as helping plants protect themselves: "When an aphid is attacked by a predator, such as ladybird, it emits an alarm pheromone. This tells other aphids to escape, and we can fool the aphids into thinking the alarm pheromone from the wheat comes from a fellow aphid."

Aphid predators are also attracted to the alarm pheromone, so the plant has two lines of defence: it repels aphids and attracts their enemies. There had been experiments spraying the pheromone onto crops, but as the compound doesn't last long in the environment, this was not a viable option. This prompted scientists to try the alternative approach of GM. The results in the lab looked extremely promising, so the team felt confident that they were heading towards a crop which would reduce the need for chemical aphid control.

Many people, however, weren't swayed by the science or by the scientists' appeal. The tension mounted when BBC Newsnight hosted a debate ten days before the protest at Rothamsted. While the presenter tried to reason with them, scientists and protestors shouted over each other in what can best be described as a nil-nil draw. Over the following few days, attempts to find common ground continued to fail.

On the day itself, the trial site was guarded by a ring of police horses, their eyes protected by plastic visors. I was there as a bystander, watching the action with a small group of onlookers sharing a 'don't destroy research' motto. Over the course of the day I spent time on both sides of the police line – talking to Take the Flour Back supporters and to scientists defending the trial.

It's fair to say that the Take the Flour Back gathering had the feeling of a family festival, with live music and organic pizza. The party atmosphere wasn't entirely lost when activists linked arms and marched up to the police cordon whilst chanting their protests. Although two men were arrested for trespassing, most of the day was entirely peaceful, and the promised numbers of protestors never materialised. Take the Flour Back put the total at 400, although the police believe numbers peaked at 200.

After the protest, the trial continued with very expensive heightened security. The field was protected by fences, CCTV, an infrared beam and security guards, and the wheat made it through to harvest. The results, however, didn't fulfil the promise of the lab.

Experimental scientists have to get used to disappointment, and that was certainly the case for this particular trial. The poor weather led to low aphid numbers, which isn't ideal for an experiment testing aphid control. However, the problem seemed to be deeper: there was no sign that the wheat was repealing the aphids. Still, this doesn't mean the project has failed. The scientists plan to continue with the work, and a future line of investigation could be to create a plant which releases the pheromone in pulses rather than continuously.

For Gia, one very positive aspect of this experience was discovering how much support and interest there was from the wider public. A Don't Destroy Research petition was signed by over 6,000 members of the public, from air traffic controllers to full-time parents. Gia found this support gave her confidence: "Although a part of society was against what we were doing, lots of people wanted us to do it and wanted to know the results. Many people we spoke to didn't have strong views on GM, but they understood the need to reduce our reliance on pesticides."

There were other signs that things could be changing, albeit very slowly. After years of stagnation of the European GM crop debate, there is now a move towards greater flexibility for individual countries. New regulations came into play in 2015 which could see transgenic crops grown in more countries. Countries can now opt-out of a GM crop approved by the EU so that it can't be grown on all or part of their territory. They must give a reason for this, for example that they don't believe the GM variety can co-exist with their current farming practices.

Hot off the marks, Scotland stated its intention to opt out of GM crop consent. When announcing the ban the country's rural affairs secretary, Richard Lochhead, spoke of consumer backlash, and said GM crops could 'damage our clean and green brand, thereby gambling with the future of our £14 billion food and drink sector'.

Amidst praise from environmental groups, many farmers and scientists lamented Scotland's decision. The Scottish Government's former chief science adviser, Professor Muffy Calder, was one of the influential scientists to speak out against the ban. She said: "It's fear of the unknown, based on some unscrupulous articles in the very early days about potential health risks which have really not been well founded and there has been no evidence ever since." A computer scientist by training, she acknowledged her lack of expertise in the area but called upon the Government to listen to the positive messages of Scottish crop scientists. Needless to say, they didn't.

Whilst many scientists spoke out as individuals, the National Farmers Union joined universities, research centres and scientific societies in sending an open letter to Richard Lochhead. The letter raised concerns that Scotland would miss out on future agricultural innovations, and claimed the decision was 'political and not based on any informed scientific assessment of risk', which is undoubtedly true. It does, however, raise the interesting question of whether decisions in a democratic society should be taken on the basis of science even if it conflicts with public opinion.

Other countries were hot on Scotland's heels to opt out, and now over half of EU member states have opted out of commercial GM cultivation.

These stories of excitement, fear and bitter disagreements are just a potted history of the GM debate. In this time a vast amount of evidence has amassed about the environmental, economic and health effects. It's no surprise that this tells a far more complex story than could ever be told by a scientist vs activist argument on the 6 o'clock news. Before we take a look at the GM industry's biggest success stories, I'm going to explore the science of genetic engineering.

Chapter 4

How to Genetically Modify a Plant

Around 10,000 years ago our ancestors made a change which would pave the way for modern civilisation: a hunter-gatherer lifestyle was gradually being replaced by agriculture. As farming became established, crop improvements began. Early farmers started to domesticate the crops we grow today simply by choosing to replant seeds from the best plants. In the millennia since, there have been some incredible demonstrations of the power of breeding. Perhaps the most remarkable is that a single plant species gave rise to broccoli, sprouts, cabbage, kale and cauliflower.

The transition from wild plants to the crops we know today was both ancient and dramatic. Native Americans transformed a short, bushy plant called teosinte into the tall crop we now know as maize (or corn, or sweetcorn, depending on where you live). The physical difference is so marked that we only recently confirmed the connection. Such drastic changes occurred long before we used lab techniques to speed up the breeding process. When archaeologists opened Ancient Egyptian tombs they found grains which look more similar to the wheat we use today than to the wild grasses which wheat originated from (although, contrary to the popular myth, these grains can't germinate).

Just by selecting the best plants to breed from, farmers ensured that wheat-yields grew fairly steadily for the last thousand years. Then, there was a sudden change. The 20th century saw an unprecedented increase in yields, as effective plant breeding was combined with mechanisation, better irrigation, fertilisers, pesticides and herbicides. The largest change came in the 1960s and 1970s, as part of what has been termed the green revolution.

The huge increase in yields which became possible in the mid-20th century was the product of increased agricultural technology along with high-yielding varieties of wheat, rice and maize. The green revolution's pioneer was Norman Borlaug, an Iowa-born agricultural scientist. His breeding programmes transformed wheat varieties, increasing both yield and disease resistance. The most striking difference in the new varieties is dwarfism. Wheat used to be tall and slender, meaning it could be flattened by the wind to become un-harvestable – a problem solved by the shorter, sturdier stems of Borlaug's new wheat varieties.

To Borlaug, food was the first component of social justice. Speaking in 1970, he said: "Food is the moral right of all who are born into this world. Yet today fifty percent of the world's population goes hungry." Borlaug spent decades working with farmers in Mexico, and honours he's received include the 1970 Nobel Peace Prize and a street named after him in Ciudad Obregon, the Mexican town where he completed his early work.

In other parts of the world, however, the green revolution still leaves people divided. It has been criticised for its use of inputs such as water, pesticides and fertilisers, including the economic impact this has on small farmers. Borlaug himself understood that the changes he contributed to aren't the whole story. In his Nobel Prize acceptance speech, he commented that the green revolution hadn't benefitted all parts of the world equally, and that staple crops such as millet and sorghum had seen little improvement. He reminded us that: "There are no miracles in agricultural production. Nor is there such a thing as a miracle variety of wheat, rice, or maize which can serve as an elixir to cure all ills of a stagnant, traditional agriculture."

From a breeder's perspective, a limitation of the modern crops which followed the green revolution is the lack of genetic variation. Just as a dressmaker wants a choice of many different fabrics, breeders want access to plants with many different characteristics. One option is to look further afield, and sometimes wild relatives of a crop have a characteristic which isn't present in the crop itself. Even when the crop and the relative wouldn't breed together in nature, breeders can often make the cross by 'rescuing' the embryos. Each cereal grain contains a tiny embryo which will grow into the plant, just as we all originated from an embryo. Grains can be dissected under a microscope to reveal the embryo, which is then germinated in a nutrient-rich medium. The downside of breeding crops with wild relatives in this way is that undesirable genes are introduced along with the desirable ones. Many of these can be removed by repeatedly breeding with the parent crop, although some will inevitably remain.

Another way to increase variation is to artificially introduce mutations. DNA can be damaged by treating seeds with chemicals, UV radiation, X-rays, gamma rays or neutrons. These mutations are often fatal, but seeds which grow can lead to plants with new characteristics. Attempts at mutagenesis go back to the 1920s, and the first crop varieties made through mutagenesis were released in 1950s.

Perhaps the most visible effect of mutagenesis is the yellow fields of oilseed rape which cover large areas of the UK each spring. Now a major component of foods such as mayonnaise and margarine, oilseed rape oil was originally poisonous. Rape had been grown as a forage crop, and production ramped up in World War 2 as an industrial oil. An intense breeding programme over the next 30 years used mutagenesis to change the oil composition so the poisonous components were reduced to a level where it was deemed safe for human consumption. The first varieties were grown in Canada in 1960, where growers named it canola, and in the USA it got the seal of approval for human consumption in 1985.

Without major public dissent, mutagenesis and crossing with related species became part of a breeder's 'traditional' tool kit. They've made great contributions to our crops – for example the ruby red colour in grapefruit is the result of gamma irradiation. Still, both techniques can introduce unwanted changes alongside the desirable ones, and they don't provide all the variation we could wish for.

New techniques to address these limitations have been made possible by our understanding of the code of life. Although Darwin and his contemporaries provided insights into evolution, they didn't know anything about the way genetic information was coded. A breakthrough in this understanding came in 1953 with the first model of the structure of DNA, from the work of Francis Crick, Rosalind Franklin, James Watson and their colleagues. Images of DNA's double helix have since become universal, and we now understand exactly how the DNA sequence codes for proteins. Crick and Watson were awarded a Nobel Prize for this discovery along with Maurice Wilkins, a colleague who Franklin had clashed with in their time working together. The Nobel Prize can be shared by a maximum of three people, and Franklin's premature death from ovarian cancer means we will never know if she would have received a Nobel Prize herself.

Before we go into laboratory techniques, it's worth taking a look at the knowledge of genetics we have built up since the structure of the double helix (DNA) was revealed. DNA is the script for life, written in exactly the same language in a human, a fruit fly and a plant. DNA comes as a double strand, with its helical backbones joined by 'base pairs'. In diagrams, these base pairs resemble the rungs of a ladder. There are four different bases, known for short as A, T, G and C, and it is the order they come in which makes the genetic code. Whilst these building blocks of DNA are identical in plants and animals, the bases come in a different sequence in different organisms. The cell is able to interpret this code, which provides a set of instructions for making proteins. These proteins perform a vast number of functions in all living cells, with hormones, enzymes and antibodies being familiar examples. Together, these proteins make something amazing, even if it is a mosquito.

A gene is a stretch of DNA which codes for a protein. Another term which keeps cropping up is the genome, which is an organism's entire DNA sequence: the complete set of instructions needed to create a living being. We have a full copy in every cell in our bodies (other than red blood cells). In the last two decades, we've seen exponential advances in genome sequencing technology; what used to take us years can now be done in hours. When 21st century techniques allowed us to sequence larger genomes we were in for some surprises. For a start, we were expecting humans to have far more genes than the approximately 20,000 we discovered, and we weren't expecting rice to have over twice as many genes as a person.

We now have an impressive understanding of many genes in human and plant genomes, although a genome is more than just the genes. In between genes there are long stretches of DNA which don't code for proteins. These were originally called 'junk

DNA', though now we know they perform a vital function in influencing which genes are copied into proteins in each cell. This is essential so that we don't, for example, produce digestive enzymes from our skin.

Our knowledge of plant genomes has led to drastic changes in conventional plant breeding. Until very recently, plant breeding has been based on using physical characteristics to select the best plant. Improving the characteristics of the crop plant and the food it produces are, after all, what we are aiming for. Still, this method has limitations – physical characteristics are affected by the environment, and they're often not visible until the plant is fully grown. They are also influenced by lots of genes, and it's not an easy task to ensure all the beneficial genes are present in a single plant.

The ideal scenario would be to select plants based on their genes, and the new technique marker-assisted selection has made this possible. In this, short fragments of DNA are used as markers to identify which plants are likely to contain the genes of interest. This can be an efficient breeding technique, partly because you don't need to wait until the plant is fully grown to see whether it has what you're after. The technique relies on knowing which DNA markers are associated with the characteristic you're interested in, and we now have 'marker maps' of a range of plants and animals used for both agriculture and forestry.

Marker-assisted selection isn't classed as genetic engineering because the technique doesn't change the DNA sequence; it just ensures that the right bits of DNA end up in your crop. If we take writing a book as an analogy, this is equivalent to marking paragraphs in different drafts to ensure they end up in the final manuscript.

Some of the barriers to its use are exactly the same as for GM techniques. For a start there is the cost, and there are also intellectual property issues because some techniques and markers are patented. Still, the technique is proving popular and it's currently being used for projects which range from selecting rice genes for a pleasant aroma to breeding silkworms with a tolerance to fluoride.

Other emerging ways to reduce the time it takes to create new varieties include a 'speed breeding' technique recently developed at the University of Queensland. This was inspired by NASA's plans for growing wheat quickly on long space missions, and uses 24hr light to allow researchers to grow six generations of wheat each year rather than the one you get in the field.

Techniques such as speed breeding and marker-assisted selection have made plant breeding faster, but they can't actually introduce new genetic variation. Genetic engineering, on the other hand, can introduce entirely new genetic sequences or make use of the variation found in other species. The terms genetic engineering, genetic modification and transgenesis are used interchangeably, and involve adding DNA into an organism's genome. As we've already seen, conventional breeding often involves modifying a plant's DNA, so some people have questioned whether genetic modification is an appropriate term – it could apply to conventional breeding too. To add to the ambiguity, some of the new techniques we will consider below mean that what we class as GM is increasingly hard to define. Regulatory systems, however, are clear about what genetic modification means.

Humans don't have the prize for the first to try genetic engineering: our ability to introduce new genes into plants was initially possible because other species were doing this already.

Sharing DNA is a common pastime for bacteria, which is one of the reasons that antibiotic resistance has spread so fast. Genes have also been shared between organisms which are important in our food systems – species that have acquired genes include the mould which gives camembert its white crust. The ability to transfer DNA to higher organisms such as plants, however, is much less common. One impressive example is that some insects, including moth species, can safely eat plants containing cyanide. It seems they gained this ability thanks to a gene they acquired from bacteria.

Gene transfer from bacteria to plants also occurs in nature, and has paved the way for genetic engineering. *Agrobacterium tumefaciens* is a common soil bacterium which infects wounded plants, causing crown gall disease. A circle of DNA in the bacteria, called a plasmid, inserts genes into the DNA of the plant. These genes cause plant cells to keep dividing, which produces a gall (a growth like a tumour). In some cases, genes introduced by the *Agrobacterium* can become permanent features of the plant DNA. Studies have shown, for example, that cultivated sweet potatoes contain DNA from *Agrobacterium*. Interestingly, these sequences aren't present in wild relatives of the sweet potato – could *Agrobacterium* have given sweet potato a characteristic which made it popular with early farmers?

Seeing no need to reinvent the wheel, scientists now use *Agrobacterium* as an effective way to introduce desired genes into plants (and it has even worked on human cells). Professor Mary-Dell Chilton is one of the pioneers of this technique, and she sees the process used for genetically engineering plants as a natural one. She said: "We learned it from nature. All we did was learn how *Agrobacterium* manages to put a gene into the plant, and we copied that process."

Her early research focussed on the basic science. Although some people were claiming that genes were transferred from *Agrobacterium* to plants, she didn't believe a word of it. She knew the theory was wrong but, to understand what was going on, her whole team worked together on a huge experiment. She said: "We dubbed it the brute force experiment because we had to work around the clock for about three days. It turned out, to my astonishment, that *Agrobacterium* was indeed guilty as charged."

Having understood this process, they worked to put it to good use. By changing a single gene, they stopped *Agrobacterium* from causing the cells to grow abnormally to produce galls. In 1983 Mary-Dell and her colleagues at Washington University announced that they had created the world's first transgenic tobacco plant. "The plant was totally useless," said Mary-Dell. "But it did show that we could do it."

With the work of many teams, engineering processes have been refined and improved. Popular options include the 'floral dip' method, where the flowers are dipped in a culture of *Agrobacterium* and go on to produce modified seeds. Alternatively, plant cells or bits of leaf can be treated with *Agrobacterium* and then grown into a whole plant using tissue culture. Tissue culture involves growing a plant in an artificial medium, which provides the nutrients and hormones needed to trigger the cells to grow. It isn't unique to genetic engineering, and its other uses include cloning plants which have favourable characteristics or orchids which are otherwise hard to breed.

Agrobacterium remains the most common way to transform a plant, and Mary-Dell has received widespread recognition. In 2013 she shared the World Food Prize, and in 2015 she was one of the National Inventors Hall of Fame's class of inductees. Now in her 70s, Mary-Dell still shares her story in lectures, along with her enthusiasm for what this technology could contribute to future food security.

One early drawback of this work was that *Agrobacterium* doesn't naturally infect many of the major crop plants, including wheat, maize and rice. Transforming cereals first became possible with the invention of the gene gun, which bombards plant cells with tungsten, silver or gold particles covered in DNA. Some of the DNA is then integrated into the plant genome, creating a genetically modified plant.

The gene gun was developed by a team at Cornell University, and was partly inspired by Professor John Sanford's fight against squirrels in his back yard (he used an air rifle to protect his bonsai trees). He joined forces with electrical engineer Professor Ed Wolf, and Nelson Allen, head machinist in the engineering lab. Ed found a Crossman air pistol on the shelf of a drug store near the university campus, and gave it to Nelson. As well as being a talented engineer, Nelson had a personal motivation; his daughter had died of leukaemia when she was 20, and he hoped the technique would bring medical advances. He modified Ed's air pistol to – and this is no more complicated than it sounds – shoot tungsten particles into whole onions, in the first of many gene gun prototypes.

In 1983, the team spent their Christmas break testing the gun and splattering themselves with exploded onion parts. Their tendency to destroy rather than transform the plant even led one researcher to attach a string to the trigger and leave the room before he fired. Gradually, they overcame the problems, using gunpowder rather than air to accelerate particles, and introducing a vacuum chamber to reduce the effect of air resistance on the particles. As they modified their gun, the team reached a point where they could shoot in a way which allowed the particles to enter the cell wall without damaging the plant.

They published their success in 1987 and, with proof the gun could be used to introduce DNA into cells, they filed for a patent. However, they were unable to attract investors so John and Ed set up a business selling the guns to other researchers. In 1990 Cornell University sold the rights to the technology to DuPont, for a higher price than Cornell had ever before earned in royalties.

Gene guns have come a long way since their invention in the 1980s: the first one was basically a BB, whereas now the commercial design uses helium as the propellant. The target is often plant cells in a petri-dish, which are then grown into a plant using tissue culture. Beyond plants, gene guns have been used in medical research, and are a potential option for delivering gene therapy to cancer patients, tackling the challenge which had motivated machinist Nelson Allen.

These techniques can introduce a synthetic DNA sequence designed in the lab or a sequence from another organism. If we go back to our book analogy, this would be equivalent to adding a new paragraph to our book, perhaps one found in another book.

When adding new genes, we also need to make sure the genetic code is translated into a protein. In the next chapter we will discuss crops which have a bacterial gene added, though adding the gene on its own isn't enough. In bacteria, the genes have their own promoter (basically the gene's 'on switch'), but this isn't recognised by plants. So the gene is often inserted along with a promoter from the cauliflower mosaic virus, a common virus which infects members of the cabbage family. This works to ensure the gene is translated into a protein.

The use of a virus initially caused campaign groups to sound the alarm, although no problems have emerged. This is perhaps not surprising seeing as we already eat crops that have virus DNA inserted in their genomes and haven't suffered any ill effects. In fact, it's unreliable to use the cauliflower mosaic virus promoter as a way to identify GM crops because it is widespread already. Just as you can't give cabbages your cold, they can't give you cauliflower mosaic virus.

Some promoters cause the protein to be produced in all the cells, though it's also possible to use a promoter which only causes protein to be produced in the green tissue and not in the seed. The use of different promoters to switch genes on in different parts of the plant is a powerful tool unique to genetic modification.

In other situations, the challenge is to switch genes off rather than switch them on. Compared to the practice of inserting new genes, a subtler approach has been used in the Arctic apple, engineered so it doesn't turn brown. Rather than adding a gene from a different species, a gene in the apple has been turned off using a technique called RNA interference. The process of creating a protein from the blueprint of a gene requires multiple types of RNA, which is very similar in structure to DNA. One role of RNA is to act as part of the regulatory process, controlling how much of each protein is produced. Small RNA molecules can cause an increase or decrease in gene activity, and the most common result is 'gene silencing' where no protein is produced. This technique was used to make Arctic apples generate less of an enzyme that apples naturally produce, and so prevent it from turning brown.

The Arctic Apple was developed by Okanagan Specialty Fruits. Neal Carter, Founder and CEO, explained: "RNAi, or RNA interference, is a naturally occurring activity in plants and animals and has been present since the beginning of time. Cells naturally use RNAi for turning genes on or off; the mechanism of RNAi is common to plants, animals, and even humans!"

RNAi is even under investigation as a treatment for cancer, silencing genes involved in cell division. Such therapies have reached early-stage clinical trials, and improved drug delivery will hopefully allow exciting new possibilities for patients.

Meanwhile, the regulatory bodies have recently concluded that Arctic apples are as safe and healthy as their conventional counterparts, and they do not present any unique environmental risks. The length of time it takes for apple trees to grow means there is a delay between approval and the apples appearing on the shelves, but the team is already moving on to new projects. They are now creating other apple varieties to add to the current Arctic Golden and Arctic Granny. The enzyme which is silenced in Arctic apples also causes browning in many other fruits and vegetables, so the team is also embarking on projects to create other non-browning crops.

Neal sees a big role for RNAi beyond his company: "Thanks to its specificity, we see RNAi as a technique that will be increasingly used. Indeed, there are already a number of other biotech crops developed through RNAi that have recently received US regulatory approval or are on the way."

Okanagan Specialty Fruits was founded by Neal and his wife Louisa, who are apple and cherry farmers. When Neal originally founded the company in 1996 he didn't envisage a 20-year journey before getting their first crop approved. However, even whilst the regulatory process dragged on, they were backed by loyal and enthusiastic investors. Neal said: "We have a great team that is small but mighty. One thing we are extremely proud of is that, not only was our team able to develop and successfully achieve approval for one of the first biotech crops with a direct consumer benefit, we did so as a grower-led company with fewer than five full-time employees – only three of whom were exclusive members of our science team."

Although the apple gained its non-browning characteristic without the need for a gene from a different species, the process still required a foreign gene to be introduced. Neal explained: "A marker gene is needed because apple transformation is, unfortunately, quite inefficient. In a single transformation, we use up to 1,000 pieces of apple leaf tissue, but typically only between one and five are successfully transformed. The marker gene we use confers resistance to the antibiotic kanamycin, which in turn allows us to check which pieces of leaf tissue were successfully transformed."

A selectable marker gene, often giving resistance to the antibiotic kanamycin, has been an essential part of the development process of many GM crops, as we saw in Chapter 2 with the story of Flavr Savr. Not all the seedlings in the experiments will have been genetically modified, and antibiotic resistance allows researchers to identify the ones which have. The antibiotic resistance genes come from the bacteria *E. coli* which is common in our guts.

Neal has no concerns about using kanamycin resistance: the gene they use is a common selectable marker for plant transformation and is sourced from naturally-occurring soil bacteria. The protein which the gene produces has been given a Generally Recognised As Safe certification from the US Food and Drug Administration (FDA). This means that, unlike Belinda and the Flavr Savr, Neal and his team didn't have to do specific tests to determine the safety of the kanamycin-resistance gene.

Despite the FDA's recognition of kanamycin's safety, critics of GM have often cited the use of antibiotic resistance markers as a cause for concern. Antibiotic resistance genes moving from plants to bacteria is a topic which has certainly attracted a great deal of scientific attention, although it appears that marker genes used at the moment haven't had any ill-effect. The European Food Safety Authority, for example, reviewed the data available for two antibiotic resistance marker genes and concluded that adverse effects on human health and the environment are unlikely. The presence of antibiotic resistance genes in some GMOs seems likely to continue, although there is a move towards alternative markers such as herbicides, or to removing the marker gene altogether once the experimental phase is over.

All these different techniques raise the interesting question of whether the definition of a GMO should be based on the techniques used to create it or the presence of a new gene from a different species. What if we used a gene gun to introduce a gene from the same species? For some objectors to GM, it may be more palatable to use genetic engineering techniques to introduce a gene from a wild relative of the crop (known as cisgenesis) or even a different variety of the crop itself (known as intragenesis). In these instances researchers are introducing exactly the same genes that would be available through conventional breeding, but they are able to do this more quickly using biotechnology. Both cisgenesis and intragenesis are still classed as GM at the moment, but some regulators are reviewing this.

In addition to the problem of tackling the regulatory system, *Agrobacterium* and gene guns come with biological drawbacks. In neither of them do you know where in the genome the DNA will be inserted, and you don't know how many copies of the new gene there will be. However, new techniques are changing this, and changing the debate along with it. These techniques are classified as genome editing, and use enzymes which insert, delete or replace bits of DNA. Genome editing allows scientists to modify precise locations on the plant's genome, so reduces the risk of unintended consequences. For decades we have been inducing random mutations which occur all over the genome, and now it's becoming possible to be much more precise.

A non-transgenic herbicide-tolerant canola (oilseed rape) went on sale in the USA in 2015, developed by the Californian company Cibus. 'SU Canola' is resistant to the herbicide sulfonylurea, and is set to reach Canada in 2017 and the global market after that.

SU Canola is specifically marketed as non-GM, and the American regulators agree – to them it is an example of mutagenesis not genetic modification. More good news for Cibus is that their crops are also considered not to be GM by the Non-GMO Project, which provides GM-free certification in the USA – anyone shopping in the USA will have noticed the butterfly Non-GMO logo. This is somewhat perverse, given that the environmental risks listed on the Non-GMO Project FAQs webpage only mention problems associated with herbicide tolerance (which we'll discuss in Chapter 6).

When I spoke with Dr Greg Gocal, Vice President of Research at Cibus, he was keen to stress that Cibus has been working with regulators since 2007, even though its products are not regulated as GM. Extensive tests have satisfied the US agencies that the product is safe and is equivalent to existing canola varieties.

To give a simple explanation of the Cibus *RTDS*™ technology, Greg sees the genome as equivalent to a copy of the Sunday Times. He said: "The technology that we're using is like a spell checker, on a molecular level, which can change one letter in the entire paper. With extremely efficient cell culture techniques and gene-editing tools we can now quickly identify these changes in spelling."

In SU Canola, just two of the 1.2 billion DNA base pairs have been changed. These changes are planned and precise, without the unknowns which accompany other techniques. This is in contrast to GM techniques which introduce known DNA sequences in unknown locations, and mutagenesis with chemicals or radiation simply makes random changes. Greg explained: "With mutagenesis you get a mutation that's interesting to you, but you also get collateral damage – maybe several hundred non-intended mutations. With our technology, we can be very precise without collateral damage."

Cibus's technology is faster than conventional breeding and genetic engineering, and Greg is excited about its potential to introduce more complex characteristics. Herbicide resistance is relatively simple – after all it has also been achieved with conventional breeding. Cibus now has its sights set on larger goals, including disease resistance, efficient water use and a better nutrition profile for oilseed rape. Their current pipeline includes a path to disease tolerance in crops. One example is their ability to target potatoes resistant to blight, which caused Ireland's Great Famine in the 19th century.

Over the next ten years, Cibus plans to develop new varieties of every major crop, and they also hope to bring their products to other world markets.

When Greg called me from his San Diego office, it was clear that he feels positive about these future possibilities for genome editing. Ten years after he joined Cibus, he is still passionate about what he does: "We have a lot of potential. Agriculture has been very exciting since the beginning of my career, and I think it is even more exciting now."

As well as its own patented techniques for genome editing, Cibus uses techniques developed elsewhere. There are an increasing number of engineered nuclease enzymes available for making precise changes to genomes, and they've caused excitement in the last few years. These enzymes can be thought of as 'molecular scissors', and the different types include zinc finger nucleases, TALENs and CRISPR.

Of all the genome editing techniques, it is CRISPR which has earned the title of breakthrough. It's based on a bacterial defence mechanism, and is used to edit DNA sequences or insert new sequences. It's fast, accurate and very cheap. CRISPR can be used to make very small changes to the plant's DNA, or to insert whole genes.

In our book analogy, gene guns and *Agrobacterium* insert paragraphs at *random* positions in the manuscript. Genome editing can insert paragraphs in *known* locations, or can make more subtle changes. Individual words can be changed or removed. Whereas the conventional breeding technique of mutagenesis changes these words at random, genome editing makes known changes in known locations.

Along with millions of dollars of investment, the awards received by the scientists involved give an idea of the importance of the discovery of CRISPR. Two of the pioneers, Professors Emmanuelle Charpentier and Jennifer Doudna, appeared in Time magazine's 2015 list of the 100 most influential people in the world.

CRISPR's many uses include therapeutics, biomedical research and crop improvement. It has been used in organisms ranging from yeast for bioenergy to (controversially) human embryos. Ultimately, biomedical researchers hope to use CRISPR in treatments for genetic disease such as sickle cell anaemia, and to alter cancer cells to make them more susceptible to chemo-therapy. CRISPR components could be injected directly into human tissues, or into cells which are removed from the body, engineered in the lab and then replaced. The ease with which CRISPR can be used to modify human DNA has rung alarm bells for some, so it may be ethics rather than technology which slows down these developments.

There will also be interesting times ahead for agriculture. CRISPR creations on the cards range from non-browning mush-rooms and disease-resistant bananas, to sterile catfish which aren't a risk to wild fish if they escape from farms. In experiments at the Roslin Institute in Scotland, CRISPR has been used to change the genomes of pigs. African swine fever is deadly to domestic pigs, yet warthogs tolerate it well, and scientists believe this is due to a slight change in a single gene. CRISPR has allowed them to edit the gene in pigs.

The EU has yet to decide whether or not it will regulate ge-nome-edited plants or animals as GMOs. The decision has been repeatedly delayed, so some European countries have drawn their own conclusions. In Sweden, Professor Stefan Jansson from Umeå University set out to convince the authorities that remov-ing a small segment of DNA doesn't count as genetic modifica-tion. As no foreign DNA is inserted, the Swedish Board of Agricul-ture interpreted the EU law to mean this wouldn't create a genetically modified organism.

The debate had come about when Stefan acquired some ge-nome edited cabbage seeds from a colleague overseas (who wished to remain anonymous). The cabbage had a photosynthe-sis gene deleted, and any plant which has this gene inactivated by mutagenesis is free to be grown without restrictions. Certainly Stefan believes that classifying his cabbage as GM would have been irrational: "It's impossible to establish if the decisive change occurred spontaneously or through human interference. In fact, it would be rather odd to have two plants that are exactly the same but one is forbidden and the other is free to cultivate without limitations."

The decision meant that he was free to grow genome-edited cabbage in his garden. He fought off the caterpillars to grow healthy cabbages, and in August 2016 he shared a meal of 'tagliatelle with CRISPRy fried vegetables' with a journalist from the Radio Sweden gardening show. This project isn't heading towards commercialisation, but the Agricultural Board's ruling has important implications for the future of genome editing in Europe.

If the current excitement is anything to go by, this powerful tool for altering DNA without encountering GMO regulations could be a game changer. As well as regulatory hurdles, genome editing could reduce some of the concerns currently associated with GMOs, such as the dominance of large corporations. The low cost of CRISPR may allow smaller companies to enter the scene, and make it more feasible to work on crops with little commercial value. As we will see in Chapter 13, however, challenges such as patents and regulations may still remain. Ultimately, the key question may be how people choose to react to genome editing in plants, animals and even humans.

Chapter 5

Insect Resistance

We've explored genetic engineering as a way to tackle a vast range of challenges, and Chapter 7 gives a flavour of just some of the creative ideas being pursued. In practice, however, only two characteristics have taken off so far. Most of GM crops currently grown are either pest resistant or herbicide resistant (or, increasingly, they're both). The advantage of pest-resistant crops is exactly as the name implies – protection from creatures which share our view that the crop is food.

Each year, pests consume enough food to feed an estimated one billion people. This loss would be even higher without the use of chemical pesticides, and so we accept the cost and environmental impacts of using them. This isn't exactly an ideal situation, especially for farmers in poorer parts of the world. For some, the cost of pesticides is simply too much for them to afford, and those who do buy pesticides often don't use appropriate safety gear. For these farmers in particular, the expense of chemicals is huge, just not as huge as the impact of losing their crop. So what if a crop could protect itself from pests without the need for chemical treatments? This has become possible for certain insects, thanks to genetic engineering and a bacteria species found in the soil.

The bacteria *Bacillus thuringiensis* (often referred to as Bt) naturally produces toxins that are poisonous to insects. Different strains of the bacteria are lethal to insects such as moths, beetles and flies, but they are all harmless to birds and mammals. We know of over 200 types of Bt toxins, specific to different insects, and so far just a few have been used to develop GM plants.

The use of Bt for pest control began long before the invention of GM technology.

In 1938, the French were the first to use Bt toxins as pest control, almost four decades after the bacteria was identified in a silkworm colony. Its use gradually spread, and Rachel Carson heralded its possibilities in her seminal book *Silent Spring*: "In eastern forests of Canada and the United States bacterial insecticides may be one important answer to the problems of such forest insects as the budworms and the gypsy moth. In 1960 both countries began field tests with a commercial preparation of *Bacillus thuringiensis*. Some of the early results have been encouraging."

Over half a century later *Bacillus thuringiensis* is now available as a pesticide under trade names such as Dipel and Thuricide, which contain Bt spores and toxins. These Bt toxins are activated when the insect eats them, causing pores to form in the lining of the gut. The insect then stops eating and dies within a couple of days. These pesticides are approved for organic growing and are easy to come by online.

They do, however, come with limitations. One problem is that when Bt is applied as a pesticide, it is quickly degraded in sunlight, so is generally ineffective within a week of spraying. Manufacturers are experimenting with ways to increase its persistence, such as inserting Bt genes into other species of bacteria. One method has triumphed: plants genetically modified to contain genes for Bt toxins. The gene for the biological pesticide Bt has now been used to create a range of genetically modified crops, including Bt cotton, Bt maize and Bt eggplant.

Persistence isn't an issue as the plant continuously makes the toxin in its leaves, and the pesticide is only in the crop and not on wild plants. Bt crops are now grown around the world and, predictably, the effects have been different in different situations.

Perhaps the most widely debated Bt crop is cotton, and it is second only to maize in terms of area planted. A major target of Bt cotton is the pink bollworm, a destructive moth native to Asia. The caterpillar is a distinctive salmon pink and chews through the cotton 'boll' before eating the seeds. It reached the USA in the 1920s and now the National Cotton Council of America estimates that it costs the American cotton industry over US$32 million each year in pest control and yield losses.

The pink bollworm remains a problem in its native Asian range. India is the country growing the largest area of Bt cotton, and much of this is grown by smallholders. These farmers are faced with huge pest problems, and in the 1990s cotton was responsible for almost half of India's pesticide use. It's perhaps not surprising that a different solution has proved so popular. Professor Vijesh Krishna from the University of Goettingen, Germany, witnessed the adoption of Bt cotton in southern India by visiting over 300 farming families between 2004 and 2008. He spoke with each farmer every two years, to discover how they were using Bt cotton and what effect this was having.

Although Bt cotton was introduced in India in 2002, it only really took off between 2005 and 2006. Part of the reason that Bt cotton was slow to spread is that initially there were only three varieties available. These weren't adapted to the local environmental conditions that farmers experienced, and it wasn't until 2005 that a wider range of varieties became available. Vijesh explained: "In the early years, the increase in the number of transgenic varieties was slow because every single Bt variety had to undergo a regulatory procedure. Uncertainty, along with a politicised public debate about the technology's risks and benefits, meant approvals took time."

Once there was enough choice in varieties, Bt cotton took off. Vijesh's second visit to farmers, in 2006, revealed drastic changes in the village communities. He asked questions on the impact of Bt cotton on yield, pesticide use and the environment, and most of the responses were very positive. "Farmers became more and more aware of the new seeds that would help them increase the cotton yield with less pesticide," he said. "Bt was a superior variety for them."

Where appropriate Bt cotton varieties are planted, Vijesh found they brought benefits. Average yield increased, and there was less variation in yield from year to year. What's even more striking is the increase in farmer income. Bt seeds are more expensive than other cotton seeds in India, even after many state governments intervened to bring the price down. Still, farmer income has increased by up to 50% in some areas.

In other parts of India, there wasn't always such good news. These initial varieties were only suitable where the crop is irrigated, yet they were introduced throughout south India even in areas prone to drought. Inevitably, these failed, and the failures were surrounded by some bad press.

Even in the areas where yields have increased, Vijesh still meets some doubters. Sometimes he has constructive discussions about corporate partnerships and responsibilities, but other times his debates are less productive: "Arguing with the anti-GM activists is really a challenge, no doubt! Some will disbelieve the on-going research – and our studies – indicating 'they are bought by Monsanto'. I then take them less seriously, 'Ya, my last month's cheque from Monsanto has just reached my pocket'. There is no point in discussing with them if they disapprove of the evidence itself."

A more legitimate fear is that GM crops will reduce farmers' choice of seeds, and local crop varieties will be replaced with a few high-yielding varieties. In the long run this could increase vulnerability, as having different varieties increases the chance that some will do well whatever the conditions. Vijesh's work has shown that this fear hasn't panned out for India's cotton – there's been no decrease in seed diversity as Bt cotton has taken over. Unlike older breeding methods, GM has the advantage that once a new characteristic has been created it can be inserted into lots of varieties with relative ease. By 2012, over 1,000 different Bt cotton varieties were planted in India.

Overall, exactly how much Bt cotton has increased yields in India is very hard to say. The effect varies from region to region and year to year. The rapid adoption of Bt cotton meant that there was only a short space of time where it was widely grown alongside conventional cotton, making comparisons difficult. Things do look positive though, and analysis of an Indian dataset collected between 2002 and 2008 suggested that Bt cotton had led to a 24% increase in cotton yield per acre. The same is true of pesticide use – whilst it seems that Bt cotton has reduced pesticide application, we will never be able to confirm exactly how much. Pesticide use was declining even before the introduction of Bt cotton, so some of the reduction we've seen may have taken place anyway.

Beyond India, the yield increase varies hugely by region, partly related to how well pests were being controlled before. Greg Jaffe, Director of Biotechnology at the Center for Science in the Public Interest (CSPI) in Washington DC, explained: "The benefit to farmers in the US has been a big reduction in pesticide use. That same Bt cotton is also grown in South Africa, and what has been the benefit there? A big increase in yield, because there the farmers didn't use pesticide, they just took the yield loss. So the same crop in a different environment has a different benefit. That's why you've got to look at things case-by-case."

In China, Bt has also reduced pesticide use, which has had benefits for farmers' health. Lower safety standards had led farmers to use pesticides in a way which frequently caused them to become ill, and we've now seen a decrease in farmer hospitalisation.

Reducing pesticide use also has the major advantage of posing a lower risk for other wildlife. Experiments on Bt plants have had good results for non-target insects: feeding tests on honeybees and their larvae, for example, have found no effect of pollen from Bt crops or of purified Bt proteins.

Although the reduction in pesticide use is a big environmental and health benefit, it has left some challenges. Bt proteins are very specific, affecting only a narrow range of species. This is good news for other wildlife, but is a problem for farmers when it leaves some pests uncontrolled. In China, for example, reduced insecticide use on Bt cotton has led to an increase in mirid bugs, which pierce plant tissue and feed on the sap. Vijesh expressed similar concerns in India, where farmers aren't always aware of the different insecticides available and how they work. Unless knowledge and information are shared alongside GM technologies, we aren't able to minimise risks while ensuring the maximum benefits are realised.

Bt cotton has been widely debated in developed countries, but consumers haven't responded in the same way as they have to GM foods – few people stop and consider the fact that cotton clothing and bedsheets are often made from Bt cotton. For anyone who is concerned about GM from a health point of view, there is a very good reason to care less about GM cotton than about GM foods – and not just that you don't have to eat it. There is no DNA or protein in cotton fibres, so cotton is exactly the same regardless of the type of plant it was grown on. The plant may be genetically modified, but the finished product isn't. For anyone concerned about GM for environmental, social or ideological reasons, this lack of interest is an oversight. Many identical issues accompany Bt cotton and Bt food crops.

The number one Bt food crop is maize, which is planted over an even larger area than cotton. One of the most common Bt genes used targets the European corn borer, and it protects against certain other moth species too. The caterpillars of this rather drab moth can have serious consequences for yield, and it is by no means confined to Europe. It arrived in the USA in the early 1900s and now infests a whole range of crops, with a particular taste for corn (maize). Perhaps appropriately, the European corn borer is also a problem in Europe, and the Bt maize variety YieldGard (MON810), which protects against the moth, is one of the few genetically modified crops approved for use in the EU. It is grown in parts of eastern and southern Europe, with Spain being the largest producer.

Although Bt maize hasn't always reduced pesticide use as much as Bt cotton has, it can still lead to reductions. Evidence in the USA indicates that pesticide use has declined in non-Bt maize as well. Studies have reported that the widespread adoption of Bt maize has led to a decline in the European corn borer, benefitting farmers in a wide area. In the first 14 years of Bt maize production, the estimated benefits to non-adopters in Illinois, Minnesota and Wisconsin alone were US$2.4 billion.

Bt genes have also been inserted into vegetable crops. In 1995, Bt potato plants were approved safe by the Environmental Protection Agency in the USA, protecting against the Colorado beetle.

The Colorado beetle has also gained the name ten-lined potato beetle because of its striking appearance, looking very much like a tiger-striped ladybird. It eats the leaves of potato plants, so the stems remain like skeletons. Genetic engineering is by no means the most imaginative tactic which has been employed to fight the Colorado beetle. During the Cold War, Warsaw pact countries ran a propaganda campaign alleging that the beetle had been deliberately introduced by the USA in an attempt to cause famines. Posters showed beetles being dropped from American planes, and offered pest-control suggestions such as drowning beetles in water or kerosene. Children in East Germany were sent out to collect beetles after school. It is indeed likely that US soldiers brought the beetle to Europe, though accidentally during World War 1, and it is still present in many countries.

Like many pests, the beetle has a long history of developing resistance to pesticides, in this case starting with DDT in the 1950s. Bt is one of the pesticides that populations of Colorado beetles are resistant to, and they're joined by some moth pests too. Much of this resistance was avoidable, and Greg Jaffe believes that a lot of industrial agriculture in the USA has had a strategy of 'use and abuse' then move on to the next technology.

The corn rootworm beetle is resistant to one of the Bt molecules in particular. "The Bt toxin that was introduced into the plant was not a high dose one," Greg said, "and so the scientists had warned us that we might get resistance to it and so we needed to be extra careful."

Careful? That's not exactly how things panned out. Greg described the reality: "Corn prices went through the roof in the mid-West and farmers grew corn on corn for four or five years in the same field. Any entomologist will tell you, independent of whether it was genetically engineered or not, that if you keep growing corn and using the same insecticides in that field you're going to put pressure for a resistant pest to develop."

An effective way to minimise the risk of resistance is to have 'refuge areas' of non-Bt crop alongside the resistant GM varieties. Non-Bt crops decrease the pressure on pests to become insect resistance, and so farmers are usually required to plant 5 – 20% of their crop area as a refuge. If a resistant insect feeds on a Bt crop it has a huge advantage, and so is likely to survive and pass on its resistance genes. Insects can lay hundreds of eggs, so if repeated generations of resistant insects find themselves on Bt crops where they have less competition from their non-resistant relatives, then the resistant population can quickly increase. However, if the offspring of the resistant insect instead find themselves in a field of non-resistant crops, then they have no advantage. Life's pretty tough to be an insect, and most will never survive to reproduce.

I discussed the issue with Monsanto's Mark Buckingham, who now works for Monsanto in the UK, not far from where he had his first agricultural job on a farm in the Cambridge Fens. He stressed the importance of responsible farm management. Even in the developed world it is hard to ensure that farmers comply with this, and it's even less feasible in parts of the world dominated by smaller farms. One solution is seed mixes containing both GM and non-GM seeds. While Mark describes these 'unstructured refuges' as a very elegant technique, he also pointed out one of the challenges: "The GM and non-GM seeds look slightly different, so some Indian farmers will pick out the non-GM seeds because they won't be so productive."

There is also some uncertainty over how effective unstructured refuges are for any species where an individual caterpillar will feed on more than one plant.

Monsanto lists limited refuge planting as a factor which may have contributed resistance of pink bollworm to its Bollgard I variety of Bt cotton in India, along with the early use of unapproved Bt cotton seed. Once resistance has developed, one option is to use more Bt genes, as resistance to one Bt toxin doesn't necessarily confer resistance to another. Bollgard I, for example, has now been replaced by Bollgard II which contains two different Bt genes. In Australia, this has already been superseded by Bollgard III.

Though many would argue that resistance is inevitable, we need to decide whether the pest control Bt provides, while it lasts, is still valuable, and whether it is a reasonable strategy to continue creating new pest-resistant varieties when old ones are no longer effective. Researchers are already pursuing different avenues for this, so Monsanto may not be the source of new varieties. Scientists at the China Agricultural University have engineered plants to contain a gene from the cotton bollworm moth itself. This gene regulates feeding behaviour, and current trials indicate this approach has potential for insect-resistant cotton.

The fate of both organic and conventional farmers spraying Bt as a pesticide is intrinsically linked with that of the farmers using Bt crops. Resistance can develop in either system, and both would have to suffer the consequences.

Vijesh believes that insect-resistant crops will become increasingly popular in developing countries: "The pest pressure is rather high here, and the insecticide chemicals are less effective – possibly due in part to less-educated farmers. In many crops, like cotton, the farm profit can be increased by about 50% by adopting Bt technology. The magnitude of the effect however depends on the magnitude of the pest problem; no surprise there!"

One crop he believes has a huge potential to benefit from Bt technology is eggplant, which is currently pesticide intensive to grow.

Eggplant has over 30 Sanskrit names; it is known as aubergine to me, as brinjal in India and as talong in the Philippines. It comes in a beautiful array of shapes and sizes and is a staple part of many Asian diets. The plant is hardy and can grow year round in many areas, but it is very susceptible to pest attack. In particular, the brinjal fruit and shoot borer can cause farmers to lose over half their yield, and generally prompts them to spray a cocktail of pesticides.

The borer is actually the caterpillar of a small and beautiful moth, its wings a striking white with chestnut patches. The grub-like caterpillars are less appealing – they eat the insides of young shoots and, worse, they bore inside the brinjals themselves and munch on the flesh. As fruit and shoot borers have become resistant to certain chemicals, farmers have moved onto alternatives which may be more damaging to health and the environment and, in some cases, are illegal. There have been attempts to crossbreed eggplants with insect-resistant wild varieties, but these haven't been successful.

Unlike the crops we've considered so far, eggplant isn't a commercial crop in Asia. Even when crops such as cotton are grown by smallholders, commercial companies are selling the seeds and the end product is often exported. It's perhaps ironic that the crop which activists have been most successful at blocking, Bt eggplant, is one developed as a public-private partnership. The private element of the partnership is Mahyco, the Indian company involved in bringing Bt cotton to market, which has a license from Monsanto to use a Bt gene in eggplant. Mahyco then sublicensed the technology royalty free to universities and public research institutes in India, Bangladesh, and the Philippines. Bt eggplant was developed with international help, including technical assistance from experts at Cornell University and funding from USAID.

The nature of this partnership means Bt eggplant has one less reason for critics to be concerned: the license agreement means farmers are free to replant seeds. The public research is introducing Bt to open pollinated varieties of eggplant. These varieties breed true, so each new generation of seeds will have the desired characteristics. In India, roughly a third of the area where brinjal is grown is planted with hybrid seeds, which can't be saved and replanted in future years (as we will discuss in Chapter 13). By donating the Bt gene for use in open-pollinated varieties which can be replanted, Mahyco isn't losing out on its own market. Mahyco focusses on higher-yielding hybrid seeds, which can be sold at approximately five times the market price of open pollinated varieties. The farmers buying the Bt open pollinated varieties were unlikely to be Mahyco's customers for these more expensive seeds.

Following the development of Bt eggplant, it underwent extensive testing. Safety tests have included assessment of the allergenicity of protein extracts, skin irritation tests in rabbits, and feeding studies with chickens, goats, cows and other animals. Criticisms include that more animals were needed for some of the feeding trials and that they should have lasted longer. Professor Govindarajan Padmanaban from the Indian Institute of Science pointed out: "You can always find flaws because these are very elaborate experiments. But more importantly, the particular gene has been around for more than a decade."

India is the world's second largest producer of eggplant, after China. Like many developing countries, it lacks a system for continuously monitoring commercialised GM crops. However, India's eggplant produces less risk of resistance than some other Bt crops because of the diverse farming system. Even though farmers don't always comply with the refuge requirements, other crops which are palatable to the fruit and shoot borer will act as a refuge.

Nine years of regulatory testing led India's Genetic Engineering Approval Committee to conclude that Bt eggplant was safe for environmental release. However, anti-GM campaigners scored a major victory when a moratorium was placed on Bt eggplant in India. In 2010, just a few months after Bt eggplant had been deemed safe for release, the Indian Ministry of Environment and Forests declared an indefinite moratorium. The decision was based largely on public meetings rather than scientific findings, a decision-making strategy which has understandably attracted criticism.

I spoke to one of the critics of the decision, economist Dr Deepthi Kolady from Cornell University. "The science says it is safe," she said. "But I think what is holding us back now is the politics."

While living in India, Deepthi developed an interest in how GM technology could contribute to food security, and how it works in certain places but not in others. She moved to Cornell University for an international perspective and her research includes exploring the social and economic impacts of Bt crops. She combines data collected for other studies with data she and her colleagues have collected by interviewing farmers with a structured questionnaire. A native of South India, Deepthi initially found conducting interviews in the Indian state of Maharashtra to be a challenge: "The language is different, the culture is different, everything is different. So I have my own local team who I trained. I was in the field for months, getting to know the people."

Although she spoke to some farmers who were opposed to the technology, in general she found that farmers were looking forward to the introduction of the Bt eggplant. "I have talked to the farmers and what they want is the maximum out of their field," she said. "They want a product they can take to market so they can get some money to feed their family and send their kids to school. The pesticide use is very high, and that's not working, and then there is this technology which is safe. I think it is our responsibility to give them the technology and educate them how to use it."

The moratorium remains in place as of 2016, though the journey for Bt eggplant is by no means over. Bangladesh has become the first country in the world to approve the commercial planting of Bt eggplant, after seven years of field and greenhouse trials. On 22nd January 2014, Agriculture Minister Matia Chowdhury officially distributed seedlings to 20 farmers from four districts. We are still waiting for results of these first small-scale plantings, but there are hopes that it will increase yield by at least 30% whilst massively reducing insecticide use. Nevertheless, while Bt protects against the worst pest, it remains to be seen how much pesticide will still be used as protection against other species.

As a food crop, eggplant is prompting a stronger reaction than cotton, and Deepthi believes the way in which cotton was introduced into India also meant it got government approval more easily. Bt cotton didn't reach farmers after years clearing a regulatory system – the first seeds were sold illegally, without government approval. Black market Bt cotton comes with risks such as no enforcement of refuge areas, so the government was faced with the choice to clamp down on the seeds or to approve and regulate them. Deepthi explained that Bt cotton was a hit with the farmers: "The pest problem at that time was very high, and farmers saw that these seeds were doing good and others were not. Then the black market spread and the government moved fast to approve Bt cotton."

Predictably, Greenpeace is amongst the critics of Bt eggplant. Zelda Soriano, Legal and Political Advisor for Greenpeace Southeast Asia, believes there's no scientific consensus about the safety of GM foods, so the field trials put consumers and the environment at risk. "It is a very widely available vegetable in the Philippines," she said. "Most people consume it on an everyday basis, so that is why it is very important for the common people to understand what is happening with our traditional eggplant varieties."

She believes the motivation behind Bt talong is profit and not food security. "Who is to gain from the introduction of Bt eggplant? Who has the patent of the gene inserted in the eggplant? I do not know of any crisis in the supply, I do not know of any challenges as to the quality of our traditional eggplant. So there is no sense, unless it is a business sense, to introduce Bt brinjal, or rather Bt talong in the Philippines."

The Philippines National Academy of Science and Technology begs to differ, expressing concerns about both yield loss and the health effects of pesticides.

Zelda has in the past spoken about the inadequacies of the regulation of GM in the Philippines, though when I asked her which extra tests she would like to see on GM plants her response was simply: "Greenpeace calls for policy shift from GM to ecological agriculture and not just for extra tests."

In December 2015, Zelda got her wish: the Philippine Supreme Court permanently banned field testing of Bt talong. The decision followed an earlier ruling by the Court of Appeals that field trials shouldn't go ahead without 'full scientific certainty' of safety to humans and the environment. The court also spoke of Bt talong as a change to 'an otherwise natural state of affairs in our ecology'. In today's world 'natural' is increasingly hard to define, and it means different things to different people. To many, high use of pesticides doesn't put the current system firmly into a classification of natural.

The irony of banning scientific testing due to an 'absence of scientific evidence' is a common story in rulings about GM. It's also common for the scientific evidence to take second place to the views of outspoken campaigners. The story in the Philippines, however, still continues. In 2016, the Philippine American Academy of Science and Engineering expressed its "strong support for the prompt resumption and continuation of research and development and field testing of Bt eggplant". Results also became available from field trials in the Philippines, showing that Bt eggplant had indeed been successful in reducing caterpillar damage. Time will tell what Bt eggplant brings to Bangladesh, and whether other countries will follow their approval.

Chapter 6

Herbicide Tolerance

Anyone who keeps a vegetable patch knows how hard it can be to keep up with the weeding, and if your vegetable patch is hundreds of acres of commercial farmland that problem will be quite a bit larger. Herbicides can reduce the resources otherwise needed for weed control, but they're not a perfect solution. You're limited by the fact that biologically your crop is very similar to a weed, so chemists have to work hard to find a chemical which kills weeds but not your harvest. Some chemicals are specific to certain plants, and it's often a case of using a few of these to kill the different types of weed. There's also a place for broad spectrum herbicides which kill a wide range of plant species, though while the crop is growing that place isn't in your field. The easiest solution would be to have a single chemical which kills all the different weeds yet leaves your crop unharmed. Enter Monsanto and Roundup Ready. In 1996 the world's first GM herbicide-resistant crop was brought to market. It was resistant to the herbicide glyphosate, and the same gene is still widely used around the world.

If you look carefully at much of the criticism of herbicide-tolerant crops, it focusses on the herbicide rather than the modification. It's therefore worth taking some time to consider the history and impact of glyphosate. Glyphosate first came on the market in 1974, under the trade name Roundup, and was the invention of Monsanto scientist John Franz. The Scientist Maga-

80

zine celebrated this in 1990, saying: "With the 20th anniversary of Earth Day just recently passed, the time was right to recognise a scientist for his discovery of an 'environmentally friendly' product."

The discovery was followed by a string of awards, including the U.S. National Medal of Technology, and Franz was very positive about his invention: "I think it's benefited mankind. It has increased fibre and food throughout the world by increasing yields and eliminating weeds."

Such praise for a herbicide seems hard to imagine now, but makes a lot more sense in the context of alternatives widely used at the time. In Silent Spring, Rachel Carson describes arsenic compounds used as herbicides: "As roadside sprays, they have cost many a farmer his cow and killed uncounted numbers of wild creatures."

Glyphosate's low toxicity to animals and the relative speed with which it is broken down in the environment have made it popular, and not just on genetically modified crops. It was initially used after the harvest to clear all the weeds left on stubble fields. This 'post-harvest pre-planting' management is still an important tool, preparing the field ready for the next crop. By volume, glyphosate is the world's most used herbicide. It is widely applied on European farms and to clear railways, and in the UK local councils spray it liberally in parks and on verges. Agriculturally, it's very popular in nations which don't grow GM crops. Figures from the UK Government show that, in 2013, approximately seven times the weight of glyphosate was sprayed on cereals as had been in 1990. In many countries it is even a common sight on the shelves of garden centres. UK regulators judge Monsanto's latest formulation of Roundup to be sufficiently safe to go without the toxicity warnings found on most household cleaning products.

Controversially, its relatively benign nature has recently been questioned. In 2015, the International Agency for Research on Cancer (part of the WHO) listed glyphosate as 'probably carcinogenic to humans'. Although the review noted that there is limited evidence for a link to cancer in humans, it cited animal studies linking it to tumours in rodents. The US Environmental Protection Agency disagrees, as does the European Food Standards agency, which states that 'no classification and labelling for carcinogenicity is warranted'. In 2016, experts from the UN's Food and Agriculture Organisation and the WHO released a statement saying that glyphosate is 'unlikely to pose a carcinogenic risk' for humans exposed to it through food. These positions have provided fuel to the fire of both sides, and has fed into the debate about herbicide-tolerant crops.

The majority of herbicide-resistant crops are indeed resistant to glyphosate, and from a purely business perspective they are a stroke of genius. Just as a company which sells printers also makes money from ink cartridges, Monsanto can profit from both seeds and the accompanying herbicide. Even now glyphosate's patent has expired, Monsanto continues to make a healthy income from new Roundup formulations. This in itself is enough to make many people condemn the technology, fearing the consequences of farmers relying so heavily on the products of one corporation. Many farmers themselves seem to have been less worried, and herbicide-tolerant crops were adopted at an astounding rate. Glyphosate-tolerant soybeans were introduced in the USA in 1996, and ten years later represented almost 90% of the soybean crop.

Farmers had been struggling with the cost of weed control in the early 1990s, especially as weeds were becoming resistant to many of the common herbicides. Herbicide-resistant crops seemed like an easy solution. They even come with added benefits such as a quicker harvest because there are fewer weeds to get in the way. For farmers, this cheaper and easier weed control is the main reason for growing herbicide-tolerant crops. There have also been reports of increased yields, though this is by no means the case in every situation.

It's not hard to find reports from universities and industry detailing the environmental and economic benefits of herbicide-resistant crops, and lots of the data looks good. For a start, the move to herbicide-resistant crops doesn't necessarily mean more herbicides are sprayed, just different herbicides. In fact, herbicide-resistant crops initially reduced the amount of herbicide used, although this reduction wasn't sustained. Still, this figure alone isn't enough to judge impact. Herbicides differ in their environmental and health impacts so we need to look further than simply how much was applied. Even though there have been concerns raised about glyphosate, many of its contemporaries are undeniably worse, and there are examples of where these are now being used less.

Herbicide-tolerant crops have made it easier to control weeds, reducing the need to rely on soil cultivation. Herbicide-tolerant crops require less tillage, bringing the environmental benefits associated with 'no-till' farming systems. Breaking up the soil, such as with a plough or harrow, is useful weed control but it can damage the topsoil and cause soil erosion. We've known for a long time that reducing tilling protects the soil – it was one of the recommendations in President Roosevelt's 1935 Soil Conservation Act. Reduced tillage is also good news for carbon emissions, both because it helps the soil to store carbon and because it reduces the fuel needed for tractor tilling.

It is perhaps ironic that replacing a mechanical technique of weed control with a herbicide has had environmental benefits –

it's another example of where the complex reality can be counter-intuitive. A recent report even estimates that the savings from fuel used in tractors in 2013 was equivalent to almost a million family cars being removed from the road, and the effect of carbon remaining in the soil could be far greater.

It won't surprise you to hear that it's not all good news though, as one man's weed is a caterpillar's castle. This couldn't be more obvious than in the case of a plant controlled by glyphosate which is called 'common milkweed' or 'butterfly flower' depending on where you are. Growing up to 2.5m tall, it is the staple diet for insects such as milkweed leaf beetles, large milkweed bugs and, most famously, the monarch butterfly.

Farmers in the USA have been trying to control milkweed for decades. It reduces yields of crops such as soybean and maize, yet many herbicides are poor at controlling it. Glyphosate is effective at controlling milkweed, and herbicide-resistant crops allowed farmers to spray it in fields. Good news for farmers, however, is not necessarily good news for the monarchs. The butterfly is facing huge threats, including loss of its wintering grounds in Mexico through logging. The loss of milkweed on agricultural land seems to be a critical factor in its alarming population declines, although to what extent this is related to GM crops is up for debate.

The National Academy of Sciences in the USA concluded that: "Studies and analyses of monarch dynamics reported as of March 2016 have not shown that suppression of milkweed by glyphosate is the cause of monarch decline. However, there is as yet no consensus among researchers that increased glyphosate use is not at all associated with decreased monarch populations."

The decline of milkweed on farmland is a pertinent reminder of a conflict between the needs of agriculture and of wildlife. Next time you walk through a field take a moment to think about what was there before the crops – forest, grassland, marsh? There's no doubt that wildlife had to make way for growing crops. Few people are too concerned that milkweed populations are reduced to make way for crops, but throw a beautiful species such as the monarch into the picture and we're reminded of exactly what we've lost. There are ways we can alter that equation: increasing yields so less land is used, or increasing farmland wildlife sometimes at the expense of yield. It doesn't eliminate the problem though – more space for crops means less space for nature. Still, you don't need to be a farmer to support the monarch and other butterflies. If you live in the USA and have a space to let milkweed grow, this is a valuable way to help the monarchs. Likewise, roadside habitats are becoming increasingly important for butterflies and other insects. These can be a good target for any anti-Roundup campaigns.

After reviewing the scientific literature and hearing from scientific experts, authors of a 2016 National Academy of Sciences report concluded that weed diversity is similar in fields planted with herbicide-resistant crops to those planted with conventional crops in the USA. Sometimes there were fewer weeds, but not fewer species in total. It is impossible to say exactly how the adoption of herbicide-tolerant crops has affected wildlife overall, though one study which keeps cropping up is the Farm Scale Evaluation (FSE). It took place in the UK from 1999 to 2003, which is telling in itself – it's hard to imagine an assessment of herbicide-tolerant crops taking place in the UK today. The lack of European data since is perhaps one of the reasons that GM sceptics often use this older study as evidence of environmental harm. Seldom, however, are the subtleties mentioned.

The publically-funded researchers studied the weeds found in fields, along with the insects and spiders. The results clearly showed that the herbicide regimes used on the GM and conventional crops affected wildlife differently, with the winners and losers depending on the type of crop. Herbicide-tolerant winter rape, for example, had the same number of weeds as conventional fields, but the weeds were different. Fewer bees and butterflies were found in these fields, and the same was true of herbicide-tolerant spring rape and sugar beet. In contrast, fields of herbicide-tolerant maize had more weeds and more bees and butterflies than their conventional counterparts. This difference is partly due to the types of herbicides used with conventional crops; those used in conventional maize were actually more powerful and persistent than glyphosate.

Fewer weeds were found in GM beet and spring rape, causing the report to conclude for these two crops: "Growing GM herbicide-tolerant beet and spring rape would mean fewer weeds and weed seeds because the broad-spectrum herbicides are more effective weedkillers than the combination of herbicides used on conventional crops."

Not only did different crops show very different results for weeds and insect numbers, there are also likely to be winners and losers within each crop. The lifestyle of different insects means some may do better than others. For example, better weed control is bad news for seed-eating insects. However, some insects could benefit from weeds being left in the field for longer – glyphosate is so effective that farmers can wait until later in the year to spray the weeds. Though this benefit might be lost if weed populations decline in the long term.

Although most of the focus is on insects, it's worth stopping to consider the weeds themselves. Since the dawn of arable farming, getting rid of 'weeds' has been necessary to make way for the crop. It's easy to forget that what constitutes a 'weed' is a very human judgement – it is simply a plant in what we deem to be the wrong place. These plants are an aspect of our biodiversity, so is it wrong to value them any less than striking animals such as butterflies? It's an interesting question, and one which can attract strong views in either direction.

Overall, the results of the study have implications which go far beyond herbicide tolerance. For a start, it shows the impossibility of being certain of environmental risk. Even years of data collection still only reveal relatively short-term impacts, and it doesn't necessarily show the effect of the crops being adopted over a much wider area. The study also can't predict herbicide regimes which farmers will adopt in reality. This kind of study is extremely valuable, but the only way to run a big enough experiment to determine the outcome of introducing the crop is to monitor the effects of a widespread adoption. It's also a prime example of judging things on a case-by-case basis. Even though they all have the same modification, the crop types gave very different results. And the results would also be different in other environments, or if farmers simply chose different herbicide regimes.

Despite these subtleties, the most common take-home message from these studies seems to be 'GM damages wildlife'. However, Professor Ottoline Leyser from the University of Cambridge interprets the results very differently: "What in fact it showed was that effective weed control is bad for wildlife. Weeds are required to support biodiversity in agricultural environments, and are currently under threat from winter planting regimes, non-GM herbicide-tolerant crops, and a range of increasingly sophisticated weed control strategies. Banning GM crops will not address this problem. It is wrong to imply that growing GM crops, rather than effective weed control, is the cause of negative effects on biodiversity."

The FSE study was a huge undertaking, and as field data is often prohibitively expensive or complex to collect, part of our assessment of the risks and benefits is through simulations. Scientists from INRA (the National Institute of Agricultural Research in France) have developed a computer model to look at how GM cropping systems might affect wildlife. Herbicide-tolerant crops can allow farmers to change their management practices – rotations often become less diverse, for example, and it's possible to sow the fields earlier. We already know that these management practices can affect the weeds found in conventional crops, causing some species of weeds to thrive while others suffer, so we would expect the same for GM varieties.

The model found that when the only change was introducing the glyphosate-tolerant variety, biodiversity indicators weren't affected. Changing management practices, however, did have an effect. For example, simplified rotations with glyphosate-tolerant maize had a negative impact. On the other hand, reduced tillage facilitated by the GM maize had a positive effect on both insects and birds. This is perhaps unsurprising given the number of reasons why ploughing and other forms of tilling can be bad for wildlife. Bird nests can be destroyed by farm machinery, for example, and tilling can reduce the abundance of food such as earthworms.

Whilst their conclusions are still tentative, these findings led to the unsurprising conclusion that different farm management regimes associated with herbicide-tolerant crops can be good or bad for wildlife. They also found variations between regions, yet another reminder of the complexity we're dealing with when making policies to benefit nature.

Another difficulty in assessing the impact of herbicide-tolerant crops is that the technology can be used in different ways. Each farmer has decisions open to them, and makes their own choices. Rothamsted Research's Professor Nigel Halford described a conversation with a farmer in the USA who was still using his old herbicide regime: "When glyphosate-resistant crops were released the farmer's herbicide supplier slashed the prices of other herbicides, which means his old routine was still economically viable. He saves glyphosate as a back-up, only spraying if blackgrass becomes a problem late in the season. So by planting glyphosate-resistant crops he is paying for peace of mind."

For many farmers, however, peace of mind is being disturbed. Glyphosate-resistant crops provide an easy strategy for weed control, and over-reliance on this one herbicide causes problems. Poor farm management has led to weeds developing resistance to glyphosate, just as our excessive use of antibiotics is leading to resistance problems in bacteria. This is a problem for weed control and, if the farmer ends up spraying a second herbicide when the first one doesn't work, it can be a problem for the environment.

These herbicide-resistant weeds are affectionately known as 'superweeds', though this is a rather confusing term as many people use it to describe a weed which gained its resistance through breeding with a GM plant. In most instances, however, it's not that the gene is jumping into weed species. Instead, chance mutations cause weeds to become resistant to herbicides. In the same way, resistance is occurring to herbicides which aren't associated with GM crops.

Although fewer species of weed are resistant to glyphosate than to other herbicides, glyphosate resistance is still a huge problem. Resistance has developed in almost 30 countries, and many of these cases are not related to GM crops. The scale of the problem is somewhat scary. Palmer pigweed, for example, can grow more than 6cm a day to reach more than 2.5m tall. The first glyphosate-resistant population was confirmed in 2005 in a cotton field in Georgia, and it's now found in at least 23 US states.

Overreliance on a single herbicide is a major cause of the problem, so one option is simply to use a wider range of herbicides. Although the lion's share of the attention for herbicide-resistant crops goes to Monsanto and Roundup, another herbicide-resistance gene on the market provides tolerance to the herbicide glufosinate. The resistance genes in LibertyLink crops were developed by Bayer CropScience, which also sells glufosinate under the non-ironic branding of 'Liberty'. Like Roundup, glufosinate kills a broad range of weeds, and also protects against some plant diseases by killing fungi and bacteria. The resistance gene originates from bacteria.

According to the Bayer website: "With LibertyLink® you get high-performing genetics coupled with better weed control than Roundup on tough-to-control weeds for high yields that deliver." However, the use of LibertyLink is still dwarfed by Roundup Ready and, inevitably, glufosinate-resistant weeds are springing up.

More recently, maize and soybean resistant to the herbicide 2,4-D have been marketed by Dow AgroSciences. Its approval was accompanied by campaigns of 'Say NO to Agent Orange Corn'. This catchy campaign title comes with an equally effective logo depicting a gasmask in front of a maize plant. It's also thoroughly misleading. Agent Orange was a mixture of two herbicides, and one of them was indeed 2,4-D. The dangers that Agent Orange posed, however, were because the other herbicide was contaminated with a dioxin which we now know poses serious health risks. In contrast, 2,4-D has since been used for decades on farms and in household weedkillers. We may now see dropping herbicide on enemy territory to be an inappropriate war tactic, but that has no bearing on whether we consider it appropriate to use 2,4-D in 21st century agriculture.

The issue of herbicide resistance has been of great interest to Australian farmer Bill Crabtree, who runs an agricultural consultancy along with his wife Monique. Also known as 'No Till Bill', he has been active in the GM debate during his global travels. Bill has seen that the farm management practices in some countries have left them vulnerable to herbicide-resistant weeds. In particular, where farmers have used glyphosate-tolerant crops in the same fields year after year, it has been a recipe for disaster. "I've watched four countries very closely: US, Canada, Brazil, Argentina. Argentina has messed it up. US has messed it up. Canada has done a great job and Brazil has done a good job," he said.

One of the reasons that Brazil has had greater success is that they grow a diversity of crop species. Bill explained how good management has benefitted Canada: "They introduced LibertyLink canola at the same time as Roundup Ready canola, and they've rotated between the different crops. Canada has also used a modest level of soil residual herbicides, similar to Australia. In 1996 the Australians were neck and neck with Canadians for taking the number one position for herbicide resistance. And now Canada has dropped right off the map – they have no significant herbicide resistance issue."

The highs and lows of glyphosate-resistant crops have certainly played out in Argentina, an early adopter of the technology. The country has reported benefits of many millions of dollars, particularly from glyphosate-tolerant soybeans. As production of GM soybean expanded, Argentinian agriculture changed. Farms became bigger, and mechanisation increased. The technology brought the benefits reported elsewhere, including a move to no-till agriculture, but these are now threatened by herbicide-resistant weeds, such as pigweed and ryegrass. Some farmers react by simply adding more glyphosate, continuing the pressure for weeds to develop resistance.

In Bill's opinion, the problem has been exacerbated by complacency from businesses and farmers. "The year they launched Roundup Ready canola I was at a field day in the US. A Monsanto guy stood up and said 'this is really exciting technology, this is going to be the end of weeds'. I chuckled and said 'this is really exciting technology, certainly very interesting, but I think you're overstepping the mark here – just last week I got a fax from Australia saying we have confirmed glyphosate-tolerant ryegrass'. Last year I met a guy who was at that field day and he said 'you really took the wind out of Monsanto's sails'."

With herbicide resistance increasing, agri-businesses are now keener to address the problem, and this was clear when I spoke to Monsanto's Mark Buckingham. "We need to protect the efficacy of herbicides such as glyphosate because they are rare," he said. "There's a strong selection pressure for weeds to develop resistance so we need better stewardship."

One of the ways which companies such as Monsanto have responded to the problem is to introduce multiple herbicide-tolerance genes into the same plant, so farmers can increase the diversity of herbicides they use. Some seeds have as many as seven different genes inserted, and these 'stacked' genes can include insect-resistance genes as well as genes for tolerance to different herbicides.

These have become extremely popular, although they aren't a substitute for more diverse weed management practices. We can't simply expect to overuse a technology and rely on the introduction of a new one. There is certainly some concern about the widespread enthusiasm for crops which are resistant to multiple herbicides.

David Mortensen, Professor of Weed and Applied Plant Ecology at Penn State University, was one of the experts who gave evidence at hearings in the US Congress to assess whether additional government oversight was needed to address the problem of herbicide-resistant weeds. He's worried that GM varieties resistant to multiple herbicides won't be used as part of effective integrated weed management so will simply create more resistance challenges. As he points out, if herbicide-resistant weed populations are addressed only with herbicides, then evolution will most likely win. After all, farmers have already reported weeds which are resistant to at least five herbicides, and some people argue that this is the greatest threat to the sustainability of our crop production systems.

Instead, David advocates better management practices to get us off the 'herbicide treadmill'. These include limited tilling to protect the soil, cover cropping (planting a crop after harvest to maintain soil fertility), crop rotation and greater attention to scouting fields for weeds so they can be controlled before they go to seed. He said: "In this case, the adapted species are driving a particular trajectory, a particular path that we're travelling down, that could significantly increase herbicide use – the kinds of herbicides that we really don't want to be using broadly across the landscape."

One issue is that herbicides and seeds sell, whereas integrated weed management is based on knowledge rather than a commodity. If integrated weed management is going to work to its full potential, it needs publically-funded research responding to local needs, and connected education programmes to ensure this knowledge reaches the farmers. As things currently stand,

David is concerned by the way that economics is directing our agricultural decisions: "It's not surprising that technologists and market-oriented companies seek marketable technological solutions to problems. Here in the States we're a market oriented culture and it strongly dictates the science we do and the way agriculture is regulated."

Given the rate of adoption of glyphosate-resistant crops, it seems fairly inevitable that stacked traits will continue to catch on. Herbicide tolerance has also now been combined with pest resistance. Monsanto developed 'Bt Roundup Ready 2 Yield', a soybean expressing both a Bt gene and a glyphosate-resistance gene, for the Brazilian market, which completed the Brazilian regulatory process in 2010. The question is whether regulators, seed companies and farmers will ensure they are used responsibly.

Responsible use of herbicides is also important in the areas where herbicide-resistant crops haven't yet led to herbicide-resistant weeds. The future of herbicide use depends partly on whether farmers change their behaviour to prevent resistance building up. Management techniques which prevent the development of resistant weeds take fewer resources than tackling resistant weeks once they occur.

This story is intimately tied to the regulation of herbicides. The first generation of herbicide-tolerant crops used a herbicide which was less environmentally damaging than many it replaced, but this won't necessarily be the case in the future. Concerns have been raised, for example, about increased spraying of synthetic auxin herbicides. These can kill plants even at very low doses, so could potentially be damaging for wild plants and nearby crops which are not resistant.

Rebecca Nesbit

It will always be impossible to be sure which herbicides would have been used, and in what quantities, if farmers were not growing herbicide-resistant crops, and it's even harder to predict which will be used in the future. However, this seems to be the key question: in what situations do herbicide-tolerant crops increase the environmental damage caused by herbicides, and when do they decrease it? And which farming practices will minimise environmental impact and increase yields?

It's also hard to determine which wider changes to farming are influenced by herbicide-tolerant crops, and which are simply part of global changes which would take place anyway. Take Argentina, for example, where there have been concerns about large-scale monocultures expanding onto new land. This is a wider (very large) problem of changing farming practices that wouldn't go away if we didn't grow GM crops, but have these crops influenced it?

Although herbicide-tolerant crops have exacerbated certain problems, none of the challenges we've considered are exclusive to GM crops. This is made abundantly clear by that fact that herbicide-tolerance isn't an exclusively GM trait, something which discussions of herbicide-tolerant crops seldom mention. The first herbicide-tolerant crop was actually introduced in 1995, for a canola variety created through mutation breeding. It was the brainchild of chemical giant BASF which developed the herbicide and worked with partners to develop the seeds. Clearfield is now available for oilseed rape, sunflower, rice, maize, wheat and lentils. Clearfield products are available in North America, South America, South Africa, Asia and Europe. Just like the GM equivalent, Clearfield seeds have extensive management recommendations such as avoiding continuous use on the same land. Predictably, they have their own resistant weed problems. Herbicide use, superweeds, corporate control – for anyone interested in the challenges of GM crops, non-GM Clearfield has them all.

In complete contrast, there are GM crops being developed which don't share these problems. In the following chapters we will consider the opportunities and challenges of a new generation of GMOs, and the bumpy roads which their creators are navigating.

Chapter 7

More Food, Fewer Inputs

Before we look at the future of GM crops, it's worth looking at the past and future of GM microorganisms. Compared to GM plants, GM microbes have a history which is both longer and quieter. The ease with which they were accepted is perhaps not surprising. There wasn't the same fear of irreversible escape into the environment, and they were tackling problems which were obvious to everyone, not just farmers. A very clear benefit came when GM bacteria were used to produce human insulin. Pig insulin had dramatically improved the life expectancy of diabetic patients in the early 20th century, but its slight difference to human insulin meant some patients experienced allergic reactions. This changed when the first synthetic human insulin was produced by GM *E. coli* bacteria in 1978, and commercial production began in 1982. The use of GM microbes also expanded beyond medicine into food. In the same year, a patent was granted for the production of the artificial sweetener aspartame, again using a GM strain of *E. coli*.

In these initial cases, the microbes produced a substance which is purified, so no traces of the microbe are present in the food or medicine. The result is an end product which isn't genetically modified; insulin made in a human pancreas is no different to insulin made by a bacteria or yeast. Microorganisms also have a long history of use in food, and genetic engineering has recently allowed us to improve them. We also now use GM microbes in products such as yoghurt. In this situation we are consuming the entire microbe, not just an extract. In 1990 the UK was the first country to grant an approval for a microbe which remains in the food: a special strain of baker's yeast engineered to make bread dough rise faster.

Today, most GM microbes used in the food industry have been altered to produce higher quantities of something they naturally make, though in some situations the microbe has an entirely new gene added. This is common in the pharmaceutical industry, and in the food sector it allows scientists to transfer genes from one microorganism into an alternative which is easier to grow and is safe for human consumption.

A whole range of GM microbes have been developed for alcohol production. We have brewers' yeast to thank for beer, but it leaves behind high-calorie starch residues. Approval was granted in 1994 for a GM yeast which can convert the starch residues into fermentable sugar. This can be used to produce a high alcohol premium beer, or a greater volume of low calorie beer.

GM yeasts are also being used in the wine industry, to improve the taste and remove some of the undesirable substances, such as histamine. We're much more familiar with histamine as something produced by our immune systems (the millions of people like me whose bodies produce excess histamine in response to allergens are far too familiar with it), but it's also present in certain foods and drinks. Levels are high in red wine, and the effects on the many people intolerant to histamine (and some who aren't) can include headaches, nausea and 'cotton wool head'.

Sadly, a complete hangover-free wine isn't around the corner. The unpleasant symptoms of the morning after have a range of causes – from dehydration to increased stomach acid – and it's hard to eliminate them all. The good news is that GM microbes are being developed to reduce the levels of various 'hangover toxins' in wine. So, depending on the acceptance and uptake of these microbes, there may be slightly less need for the aspirin, Berocca and bouillon soup.

Microbes also allow us to be more creative with our dairy products. In cheese making, the milk needs to be transformed into solid 'curds' which are separated from the whey, then processed and matured. In almost all cheeses this is done by adding rennet, which contains the enzyme chymosin. Traditionally, rennet has come from the stomach of slaughtered calves. This is a problem for vegetarians, and GM provided a solution. In the late 1980s scientists transferred the cow gene for chymosin into microorganisms. Grown in culture, these can produce much higher concentrations of the enzyme than can be extracted from calves. Today the majority of the hard cheese made in the UK and USA uses chymosin from GM microbes rather than from animal rennet.

Many of the arguments against GM simply don't hold for rennet produced by GM microbes. The enzyme is purified, so no traces of GM material remain in the product – in fact the enzyme itself breaks down as the cheese matures. This eliminates health concerns. It is also cultured in a controlled environment so there's no risk to ecosystems.

Still, not everyone's happy. The Soil Association has ruled that organic products must not use GMOs or their derivatives, including enzymes. Its website states: "Every cheese manufacturer must submit a GM declaration from their rennet supplier that the material is not derived from a GM organism."

The Soil Association does recognise that GM microbes and GM crops are very different, and so it commissioned some research. Leading this was their Standards Project Manager Sarah Compson, who explained: "Our research indicated that there are circumstances where it would be difficult to maintain objections to genetically-modified microbes on environmental grounds alone. Disposal techniques are thorough and effective and even if accidental spills occurred, the genetic constructions are such that the microorganisms would not be viable. However, whilst there are some examples of highly rigorous safety protocols and environmental protection, these are by no means universal."

She also pointed out that, although no consumer attitude surveys had been done, the Soil Association expects that genetically modified microbes would be at odds with consumers' expectation of organic foods. Ultimately, its certification scheme is a service to consumers.

It is now possible to source suitable enzymes from non-GM bacteria and yeasts. This a relief for vegetarians keen to eat organically, though rennet from these sources tends to produce other reactions which affect the taste of the cheese.

Elsewhere, there is an even more ambitious plan for microbes in cheese production. Every vegan and aspiring vegan has weighed up the taste and nutrition of animal products against the ethics of producing them, but only a few have considered how they can have both. In California, a small group of citizen scientists have come up with an ingenious plan for delicious vegan cheese. They have created synthetic DNA based on cow genes for milk-proteins, and introduced the synthetic genes into baker's yeast.

Nobody pays these dedicated 'biohackers' for their work, they get money for their ongoing experiments through crowd-funded donations. Instead of profit, their motivations are tackling future food scarcity, and a belief that "using cows as food production machines is environmentally irresponsible".

The really interesting question is whether we can get used to the idea of Real Vegan Cheese coming from a lab. Over recent decades, we have accepted that our cheese comes from cows kept in small pens and transported long distances in crowded trucks. Given the benefits to animal welfare and the environment, I hope society can now accept this new way of making cheese.

Back in the world of plants, a major motivation for genetic engineering is the prevention of disease. Historically, plant diseases have been responsible for disasters such as the Irish potato famine, and they still reduce yields and cause economic losses of billions of dollars. They can be tackled with management techniques such as burning infected plants or fumigating soil, and chemical treatments such as synthetic fungicides or, if you're an organic farmer, naturally-occurring chemicals such as copper sulfate or peracetic acid. In some cases, however, these strategies aren't enough.

In the 1990s, the papaya ringspot virus (PRSV) threatened to eliminate Hawaii's commercial papaya production. Over two decades later, papayas are still widely grown, and it is largely thanks to the passion of Professor Dennis Gonsalves. Dennis was born and raised on a sugarcane plantation in Hawaii. As a child, he was held back by a stutter. Even by the time he graduated with a degree in horticulture from the University of Hawaii, he was resolutely avoiding public speaking rather than seeking help. His first job was as a plant pathology technician, and he was tasked with identifying a papaya disease found on one of the islands. He identified it as tomato spotted wilt virus, and knew that plant pathology was his calling.

He embarked on a research career which took him to mainland USA, and it was just before the end of his PhD that he was cured of his stuttering. He had an interview for an assistant professorship and was determined to give a clear seminar. Hours of practice in front of the bedroom mirror combined with the confidence he'd gained through his work led to a flawless presentation, a job offer and the end of his stuttering.

When the PRSV virus, carried by aphids, destroyed papaya farming on Oahu (the island which is home to Honolulu) in the 1950s, the growers simply moved islands. But it seemed only a matter of time before the virus caught up with them and the industry would once again be under threat. In the late 1970s, while he was working as a plant pathologist at Cornell University, Dennis decided to pre-empt the inevitable spread of the virus.

The first line of attack was to work on the virus rather than the plant. The team tried to develop a mild strain of the virus, using nitrous acid to induce mutations. The hope was that the mild strain would protect the plants against more virulent strains, just as cowpox protected milkmaids against smallpox. However, even after years of work, the results didn't look promising.

Dennis said: "I asked myself the question, 'Now tell me, what have you really accomplished?' My conclusion was that the approach was not good enough for sustainable control in Hawaii, and I elected to change control strategies. Giving up your favourite toy is not easy, but sometimes you need to do it to make progress."

It was time to change tack, and he realised molecular biology was the way to go. This was the very early days of plant biotechnology – the first genetically modified plant was created in 1983 – and it took years to create the first strain of resistant papaya. The virus hadn't yet reached the main papaya growing area, but its appearance close-by added to the urgency. In the late eighties, Dennis and his colleagues used a gene gun to introduce a gene from the virus into the plant.

They had used a 'pathogen-derived resistance' approach which had been shown to work against the tobacco mosaic virus. The theory was that plants containing a gene of the pathogen would become resistant to that particular pathogen, a system which Dennis' colleague Professor Richard Manshardt described as somewhat analogous to immunisation. Since people eat virus-infected fruit with no ill effects, there weren't concerns for human health.

Richard described some of the early progress: "Everyone laughed at it initially. They said 'shooting genes into cells, that's ridiculous!' But actually it worked."

Papaya plants containing the PRSV coat protein gene did indeed prove highly resistant. In 1991 Dennis and his team announced the creation of a red-fleshed 'Sunset' papaya resistant to the Hawaiian strain of PRSV in greenhouse tests. The timing was pertinent. In 1992, papaya ringspot virus (PRSV) was discovered in the Puna district of the Big Island where 95% of Hawaii's papaya was being grown. Over the next couple of years, measures to control the virus failed; farmers abandoned infected fields, leaving a reservoir of disease. The Hawaiian Department of Agriculture declared PRSV uncontrollable by standard methods in 1994.

Meanwhile, there was rapid progress with Dennis and Richard's project. The Animal Plant Health Inspection Service (APHIS) of USDA approved a field trial at the University of Hawaii which demonstrated the strength of the modified papaya's disease resistance. Now they needed to create a papaya variety which would be acceptable to growers. So far, the work had been done on red-fleshed papayas, although the local choice was yellow fruit. They crossed their red-fleshed GM variety, SunUp, with the local yellow variety to develop the orange-fleshed Rainbow. These were field tested in 1995.

The outcome was high-yielding plants producing fruit whose taste, colour and shipping characteristics passed the test. Dennis described the results of the trial: "The growth differences between the transgenic and non-transgenic trees were remarkable; transgenic plants grew vigorously, with dark green leaves and full fruit columns, whereas non-transgenic plants were stunted, with yellow and mosaic leaves and very sparse fruit columns."

After extensive testing of safety and nutrition, the fruit received FDA approval in 1997. Alongside regulatory approval, there were intellectual property issues to navigate. Ultimately four different companies had ownership of the technologies they had used to create the papaya. This included one which Washington University had licensed to Monsanto. Eventually the licenses came through, allowing Rainbow seeds to be distributed for free to Hawaii's growers.

On 1st May 1998 the first commercial seeds were released. Richard described what happened next: "The immediate result of introductions wasn't an overnight success. It took time to make enough seed for all the growers, so it was a couple of years before the seed was widely distributed."

By 2001 transgenic papaya made up about 45% of the plantings in Puna, by 2009 this was up to 77%, and in 2015 it was 85%. Not only could these varieties be grown without concern for the disease, they also reduced the level of virus in the area. This made growing conventional papayas more feasible, and allowed growers to keep the lucrative Japanese market alive, even though at the time Japan didn't allow the sale of transgenic fruits.

It was 2011 before Japanese authorities completed their safety assessment and approved the virus-resistant papayas. Dennis said: "Many people said we would not succeed. In fact, some of my friends would periodically ask me how things were coming with the Japan project; I would say fine and that we were getting closer, and they would jokingly say 'Dennis, you told me that last year!' But in the end, I had the last laugh."

Papayas were the first GM fruit to enter the Japanese market in unprocessed form, and now Japan is one of the world's largest per capita importers of food and feed produced using GM technology, with over 200 foods and food additives approved for import.

Overall, the team are extremely happy with the way things have panned out. Dennis said: "I think it is safe to say that the transgenic papaya saved Hawaii's papaya industry. Furthermore, it was the first public-sector-developed transgenic crop to be deregulated and commercialised in the United States. One might say it was a poor man's transgenic project."

There are, of course, some dissenters. Richard is conscious that fear of GMOs serves organic papaya growers' interests, and believes that their opposition is also fuelled by fear that cross-pollination by Rainbow plants will render their fruit unmarketable as organic, or that they will be sued by industry for growing GM crops they hadn't purchased. He describes the reality: "I don't know of anyone in Hawaii who has lost their certification due to accidental presence of GM crops, and there's no history of anyone being prosecuted."

However, without careful management it is perfectly possible that GM papayas will end up in organic fields. Tests performed by the scientists behind the project show that Rainbow flowers can indeed pollinate non-GM papaya varieties, although thankfully the nature of the papay tree makes this easy to prevent. The commercial Hawaiian papaya industry is based on production of fruit from hermaphrodite plants. As these trees are both male and female, the flowers pollinate themselves before the petals open. Richard explained: "No matter what kind of pollen is about once the flowers have opened, it doesn't matter, they will already have been pollinated."

Cross-pollination does occur when female plants are grown. Seed mixes have roughly one female plant per two hermaphrodites, and commercial growers will remove all the female trees. Even if cross-pollination was common, organic growers could be certain of not selling GM fruit. The organic fruit itself develops from the non-transgenic mother plant, so it won't be GM even if the flower was fertilised with GM pollen (only the seed will be GM). And to ensure the next generation of plants is also GM free, you place paper bags around flowers so no pollen comes in from elsewhere. With each flower containing hundreds of seeds, you only need to do this with a few bags. This is common practice in other situations where genetic purity is maintained by reproductive isolation, such as preserving the white kernels of America's Silver Queen sweetcorn variety.

Still, there was initial resistance to the technology among certain grower groups. Richard said: "Even though the industry was desperate for some kind of relief from the disease, some growers used this issue to leverage member support in on-going power struggles over leadership and assessments for advertising. Except for these contests within the industry itself, including a couple of meetings which were pretty divisive, there wasn't any notable consumer resistance. Then three or four years after the initial release there was the first activist reaction, and it's been pretty constant ever since."

Consumers are well known not to put their money where their mouth is, and the surprising results of a 2011 study show that the minority who shout the loudest don't necessarily reflect the views of the majority. In six Hawaiian grocery stores papayas labelled as GMOs were placed next to unlabelled fruits. The labels actually increased sales, perhaps because a labelled product is perceived to be of higher quality, or because consumers were interested to try the GM variety.

In fact, the GM papaya may be an easy target in connection with a different battle. Hawaii's year-round warm weather and fertile soil had made the islands a popular research spot for companies developing new GM crops. This corporate involvement seems to be connected to the rise of anti-biotech sentiments, and activists have turned their attention to the papaya.

In 2011 the Maui News reported: "Thousands of papaya trees were chopped down on 10 acres of Big Island farmland under the cover of night. Hawaii County police said the destruction appeared to be done with a machete, but there are no leads and few clues beyond the tree stumps and all the fruit left to rot. A growing theory among farmers is that the attack was an act of eco-terrorism, a violent protest against the biotechnology used in growing papayas here."

Unfortunately, the water is muddied by misinformation and unsubstantiated claims. Vocal opponents include GM Free Hawaii and Babes Against Biotech (who must have the prize for the most unusual fund-raising tactic – they have produced a calendar with the cover image of a 'Babe' holding a papaya, wearing a bikini and a gas mask).

Organic farmer Melanie Bondera is one of the leaders. Encouraging fear of GMOs is probably good for her profits, but it is clear she also has a passion for the environment. Her concerns include the origin of GM crops making native Hawaiians "tired of outsiders throwing their weight around". She is also worried about the unknown implications on health and the environment: "We're going to find a very long future of discovering the problems with this product."

Her campaigns have had some success. In 2013, County Council members signed into law a bill that severely restricts or, in most cases, bans the production of GM crops on the Big Island. GM papaya was exempt from the ban, though growers have to register where they grow GM papaya, document their pest control practices, and pay a fee. The fee is perhaps somewhat ironic, given that much of the backlash is against companies profiting from GM technology, yet the GM papaya had started out as free for growers.

The process of collecting evidence which led to the bill involved hearing from stakeholders such as Melanie Bondera. Academic scientists also had their say, and Dr Susan Miyasaka, an agronomist at the University of Hawaii, ended her presentation by saying: "People are allowed to have their own opinions, but not their own facts."

It is unsurprising, for example, that people have different opinions on which risks are worth taking. That doesn't mean it's appropriate to either elevate or downplay evidence about the likelihood of a particular problem. A valid opinion might be 'I believe there should be more safety testing'. However, that doesn't make a statement such as 'infertility, immune dysfunction and faulty insulin regulation are serious health risks of consuming GMOs' a valid fact.

A similar threat faces America's citrus trees, and the billions of dollars they contribute to the economy. Citrus greening disease made its first US appearance in Florida in 2005 and has now killed millions of citrus plants, infecting oranges, lemons and grapefruits.

This time the culprit is a bacteria spread by a jumping plant-louse, the Asian citrus psyllid. The disease blocks the flow of nutrients, causing the leaves to turn yellow and the fruit to remain small, hard and sour. Most infected trees die within a few years. Commercial orchards are tackling the problem with pesticides to kill the psyllids, but this isn't a viable option for many smaller orchards, and is by no means environmentally desirable. As a result, Florida homeowners were advised to remove all citrus trees in order to protect the industry.

Another option to control the Asian citrus psyllid is to intro-duce a natural enemy. California's Department of Food and Agriculture has been releasing tiny parasitic wasps native to Pakistan. At best, however, this labour intensive strategy cuts psyllid populations by a third.

In 2010, a report from the US National Research Council con-cluded that the most powerful long-term management tool is likely to be the cultivation of citrus trees resistant to the bacteria that cause citrus greening and to the Asian citrus psyllid. To try and create a resistant tree, Texan Professor Erik Mirkov introduced genes from spinach which protect against the bacteria. These genes were chosen because spinach is already consumed safely, and the team hopes it will therefore be more favourably received by consumers. Experimentally, the trees appear to be resistant to the disease, and in 2015 the U.S. Environmental Protection Agency approved an application for field trials. Commercial release may be a few years off, but this will be an interesting story to follow.

Work on disease resistance isn't limited to trees, and much of the early research took place on the potato. Potatoes have the advantage that they are sterile and grown from tubers rather than seeds. This means there isn't the same potential problem of cross-pollination, although it hasn't been enough to make them popular.

In the 1990s, Monsanto developed potato varieties resistant to the leafroll virus, which causes severe economic losses for American farmers. The varieties were commercialised in 1998 and were appreciated by farmers. However, the potatoes weren't so well received by many consumers, and a campaign was launched against them. As a result, McDonald's declared it would not sell fries from GM potatoes. This is a sentiment the company has recently reiterated, even when its major supplier developed a potato which bruises less and produces less acrylamide, a chemical compound that some studies have linked to cancer.

Still, all is not lost for the disease-resistant potato. British scientists have created a variety which is resistant to late blight. In the 1840s the disease was responsible for the Irish potato famine, and in 21st century America it can lead to financial ruin, farmer suicide and the application of large volumes of fungicides. Farmers in infected areas often spray more than ten times a year to tackle the disease, so scientists at the UK's John Innes Centre have created a transgenic blight-resistant variety. The blight-resistance gene came from a South American wild relative of the potato, and a small-scale trial over three years found that the GM variety increased yields. The trial predictably caused controversy, although a single protest in 2011 only attracted around 60 activists. One criticism was the cost to taxpayers, to which scientists responded by pointing out that UK farmers spend an average of £60m a year controlling blight, with worldwide losses of approximately £3.5bn.

Overall, disease-resistant plants have the potential to reduce the use of both fungicides and pesticides. However, they do raise some specific environmental questions. For example, are there situations where we risk changing the community of soil bacteria? There's much left to learn about the bacteria below our feet, yet these are essential for soil fertility. Plants can release defence chemicals from their roots, and it is important to show that any genetic modification doesn't negatively affect soil bacteria and fungi. This kind of question needs to be answered in different situations and for different crops, but recent studies have tackled it for wheat modified to resist yellow mosaic virus.

Just like in the virus-resistant papaya, the inserted gene came from the virus itself. A two year study by the Chinese Academy of Agricultural Sciences showed no adverse effect on soil microbes. The Academy is one of the world's largest investors in biotech wheat, and they have other diseases in their resistance pipeline. We may have interesting times ahead for disease resistance.

There's also biotech potential for pest control, going beyond the limited species affected by Bt proteins. When you think of pests, it tends to be caterpillars which spring to mind, not the tiny roundworms which inhabit the top layer of our soil. They are amongst the most numerous organisms on Earth and can be herbivorous, carnivorous or parasitic. As parasites of slugs and insects, some nematodes are seen as gardeners' friends, while those which parasitise plants are less popular. The value of the annual crop loss from nematodes has been estimated as US$118 billion globally. We do have nematicides to control them, but these are environmentally harsh as well as not fully effective. Crops resistant to nematodes have been through field trials, and the biggest constraints seem to be market acceptance rather than technological.

These crops developed for enhanced pest control have very simple modifications, as do most of the plants we've considered so far. Can we achieve the same kind of advances when many more genes are involved? As first-generation GM crops caught on, more people began to research the harder problem of tolerance to environmental stresses. The stakes are high – we're currently losing yield due to limited water and minerals, or we're adding them at an environmental and economic cost. In many developing countries 90% of freshwater withdrawals are used for agriculture. The task, however, is far harder than introducing pest or herbicide resistance. Plants' reactions to shortages of water, nitrogen and phosphate are complex and so conventional breeding hasn't always made great progress.

Biological complexity is one of the reasons that GM sceptics believe biotechnology can't tackle these problems. Environmental campaigner Vicki Hird said: "More knowledge and skills-based systems – farmer-centric agronomic approaches such as agroecology – and possibly other genomic technologies publicly owned and developed, are more suited to the complex challenges ahead. I've yet to see evidence that drought resistance or nitrogen fixating is anything but a lab dream. There are so many factors involved in these complex biological mechanisms – herbicide resistance or Bt toxins are easy."

Despite the complexity we're dealing with, many scientists haven't been deterred. In fact, they have risen to some of the toughest challenges.

One very common limiting factor for crops is nitrogen, and correcting this with nitrogen fertilisers is a major environmental problem. Every time we take a breath, our lungs are filled with air that is approximately 78% nitrogen, yet to most plants this form of nitrogen is useless. The exception is legumes, such as peas and beans, which have a relationship with bacteria allowing them to turn nitrogen from the atmosphere into a useable form. In some situations, this gives us the powerful option of crop rotations which use legumes to enrich the soil with useable nitrogen. However, this isn't generally the avenue we go down.

Instead we react nitrogen in the air with natural gas to make fertilisers, in an energy-intensive process. The environmental concerns don't end with the energy input – once it's added to the field, the fertiliser can be washed off into rivers and lakes. Here excess nitrogen can cause algal blooms, disrupting the natural ecosystem and reducing oxygen levels in the water. It's a problem wherever nitrogen fertiliser is used, but particularly shocking in countries such as China with lower environmental safety standards.

This environmental damage, however, wasn't the reason that the Bill & Melinda Gates Foundation chose to fund projects tackling the nitrogen problem; it was the wellbeing of smallholder farmers. Smallholders make up about 75% of the world's poorest people, those who live on less than a dollar a day. Most don't have access to nitrogen fertilisers, and they deplete the soil nutrients by growing crops such as maize on the same land year after year. The result is crops that don't come close to their potential yields.

In 2011, the effect on yield prompted the Gates Foundation to bring 20 scientists from around the world to Seattle for a convening on nitrogen. From this they commissioned two projects with the bold aim of creating cereal crops that can fix their own nitrogen. Unsurprisingly for such a huge step in biology, both projects use genetic modification. One is transferring genes from bacteria into cereals and the other is transferring genes from legumes. With the 'prototype' of legumes proving plants can fix nitrogen, a team from the UK's John Innes Centre have taken this as their starting point.

The project is led by Professor Giles Oldroyd, who believes this could be a game changer both in reducing dependence on nitrogen fertiliser in the developed world and in raising yield elsewhere. He explained the potential: "If you want to get to a sustainable agricultural system that allows equitable food production across the planet, then you can't do it without tackling nitrogen. Putting the capability for nitrogen fixation into cereals has the potential to be truly transformative of yield production. I would say if we had nitrogen fixing cereals now it would change the nature of the GM debate. It really is the next generation of GM crops."

The challenge is complexity. Whereas other GM crops might have a single gene inserted, for nitrogen fixation you need entire biological pathways. This means that Giles and his team have their work cut out, and are only at the early stages: "We have to put in a lot of new technology, push the boundaries of plant synthetic biology in order to achieve it. However, I'm hoping that as we go along that route we can start discovering something useful. I would anticipate that the engineering process is going to track a gradually increasing level of nitrogen fixation. We're not going to have this all or nothing process where you suddenly get nitrogen fixation in cereals and it's as good as a pea plant. And even modest levels of nitrogen fixation would be transformative for small holder farming systems, just because they're starting from such a low level."

Closer to home, Giles has been the target of criticism by anti-GM campaigners. When the project was launched in 2012, Friends of the Earth Food and Farming Campaigner Kirtana Chandra-sekaran was quoted saying: "GM crops have over-promised and under-delivered – this grant would be better spent helping developing countries build on their traditional knowledge to develop diverse and resilient farming systems that meet local needs."

Giles hoped to talk this through personally, rather than through the medium of news articles. He said: "I emailed her on a couple of occasions and asked her directly 'what do you mean by traditional farming practices, can I have some examples?'. I really wanted to open a discussion on this. I would say I'm pretty open minded here – I'm doing this because I think it's for the betterment of these farming systems, not to lock them into some corporate system. She just never responded to me. In my view traditional farming in Africa is what they are already doing, which is using bad seed, just spreading it over the field and hoping for the best. I don't know what aspect of that she would like to develop or invest in."

I too asked Kirtana for her ideas and, like Giles, I received no reply. Still, she is right that there are other options that are worthy of consideration. Current organic alternatives to nitrogen fertiliser include adding manure to the crops. Many smallholders already do this, though most have few domestic animals so the little manure they have is often human waste. Growing legumes is another option, and organic farmers in the developed world have success growing clover as part of a rotation system. For conventional farmers, the extra yield they get from never having to plant clover is worth the money they spend on fertilisers. I will consider agroecology practices in more detail in Chapter 19, although Giles sees the idea that agroecological systems could replace all the technologies of the developing world as ridiculous.

He does, however, have more understanding of those who question whether the US$10 million of funding he's received will lead to success. He said: "It is a high risk project, and I fully accept that. It might take 50 or 100 years, I don't know. But, from my view, it has the potential to be so transformative that you have to have some of these high risk programmes."

Whilst tax-payers would perhaps be reluctant to foot the bill, this kind of high risk project is something the Gates Foundation is keen on funding. As Bill and Melinda Gates say on their website: "We think an essential role of philanthropy is to make bets on promising solutions that governments and businesses can't afford to make."

Along with nutrients, water availability can be a major limit to productivity. The Intergovernmental Panel on Climate Change's prediction of an increase in extreme weather events is not good news for crop yields or the reliability of our food supply. The severe droughts experienced by two major food exporters, Australia in 2006–2007 and Russia in 2010, may be warnings of what's to come. The food price spike of 2008, which I noticed when I bought my bread, was a reminder to policy makers that the food security of the last few decades can't be taken for granted. And it isn't just yield which can be affected by environmental stress, it is also food quality, both the flavour and nutritional value.

Modern crop varieties are thirsty, although recently conventional breeding has produced crop varieties which are slightly better able to cope with limited water. We also have contributions from genetic modification, and in 2013 Monsanto's DroughtGard was approved for commercial growth. It has been met with mixed levels of enthusiasm. The crop doesn't survive severe droughts and, whilst it does produce better yields than conventional equivalents in dry years, the increase in production is fairly modest. Even Monsanto is managing expectations. As spokesperson Danielle Stuart said: "This isn't a product that we're expecting to grow in the desert."

The area of the USA planted with DroughtGard is increasing, but the more exciting work is going on elsewhere. Some of the most serious droughts hit African nations, where maize is the most widely grown staple crop. At least 300 million Africans depend on it as their main food source, which leaves them vulnerable to fluctuations in maize yield. In some years, combinations of drought and insect damage can cause complete crop failures. Genetic techniques have the potential to tackle these two challenges simultaneously.

The African Agricultural Technology Foundation is coordinating the Water Efficient Maize for Africa (WEMA) project to develop maize which is both drought-tolerant and insect-resistant. They are using conventional breeding as well as biotechnology, and they aim to offer these varieties royalty-free to smallholders in Sub-Saharan Africa. Smallholders aren't the most lucrative market, so this is a public-private partnership, with partners including African research institutes, the United States International Agency for Development and Monsanto. WEMA is a philanthropic project for Monsanto and, in addition to staff time, it is donating genetically modified traits of insect resistance and water-use efficiency royalty-free along with proprietary maize varieties.

WEMA has already released DroughtTEGO, a maize variety developed using conventional methods, and feedback from farmers has been positive. However, they believe that biotechnology will allow them to offer even better varieties. Controlled field trials are currently taking place in agricultural institutes in Kenya, Uganda and South Africa. Assuming there are no unexpected regulatory hurdles, the project hopes to have transgenic maize varieties available towards the end of the decade.

At the other extreme, climate change is also likely to be responsible for increased flooding. I grew up in Tewkesbury, a small English town known mostly for the time it spends under water, so this is a topic close to my heart. It is also already a major issue for food security. In India and Bangladesh, 4 million tonnes of rice are lost to flooding each year – that's enough to feed 30 million people. The problem occurs because the high-yielding rice varieties which farmers like to grow aren't very good at surviving submergence. Older varieties are much better at this, and amazingly it seems that just one gene is largely responsible. Whereas previous breeding programmes had little success in crossing the flood-tolerant varieties with modern ones, once the gene had been identified it was introduced into modern varieties. They used marker-assisted selection, a technique we discussed in Chapter 4, which isn't classed as genetic modification. This avoided the expensive regulatory hurdles and meant the technology quickly reached farmers. In 2011, 1 million farmers planted Sub1 rice, which can survive submergence of up to two weeks. Just two years later, this number had grown to 4 million.

GM approaches could open up new options to produce rice which could stay submerged for longer in more extreme conditions. Researchers have identified other genes involved in flood tolerance which they hope will be useful to create 'Sub1 plus' rice varieties. One of the scientists leading this work is Professor Pamela Ronald from the University of California, Davis. She has been working on Sub1 rice since 1996 and is passionate about using all appropriate technologies to advance sustainable agriculture and food security. She said: "This includes planting crops developed through diverse modern breeding technologies – marker-assisted breeding, genetic engineering, genome editing, hybridisation – and cultivating those crops using ecologically-based farming practices."

The success of Sub1 rice created through non-GM techniques raises interesting questions. Does it mean that we should stick at this and avoid GM techniques altogether? Or is it a sign that concerns over GM are rather arbitrary – using another lab technique to alter a plant's DNA to confer flood tolerance is clearly acceptable, so what is different about GM?

Innovations such as Sub1 rice have the potential to tackle some of the challenges which cause crops to be less productive than they theoretically could be in ideal circumstances. Pests, disease, water shortage, mineral deficiencies – there is a long list of limiting factors to tackle, bringing yields closer to the maximum 'potential yield'. All the cases we've looked at so far have been focussed on closing the gap between the actual yield and the potential yield. Is it time to aim even higher? Can we raise the bar and create crops which have a higher yield potential?

Professor Louise Fresco, who studies international sustainable development at Wageningen University in the Netherlands, is an advocate of this: "There are two ways which genetics can help when it comes to agriculture. One is to bridge the yield gap when you are still very far from breaching that – for example, that's the case in Africa. The other one is to actually move the biological maximum yield, to go beyond what we ever imagined was possible."

During the 20th century we increased yield potential by breeding crops which concentrate more of their energy in the parts we can eat. It's been an incredible strategy for increasing yields, but we're potentially near the upper limit for this. Next, some scientists have turned their attention to photosynthesis. Photosynthesis, the process of carbon dioxide and water into glucose and oxygen, is the pillar of life on our planet. Still, plants might get a 'could do better' comment on their report card. If plants could convert more sunlight into sugar, that would make this energy available to people. There are lots of potential angles we could come from, such as increasing light capture or the uptake of carbon dioxide. Plants can't efficiently use the high light intensities found in the middle of the day, for example, so that energy goes to waste.

One possible target for improving photosynthesis is the enzyme RuBisCO. It is the most abundant enzyme on Earth, but works relatively slowly. It evolved in very different atmospheric conditions (at a time when atmospheric CO_2 levels were much lower), and researchers believe that modifications could allow it to work faster in today's atmosphere. Creating a faster RuBisCO is no mean feat, and will probably rely on an increased understanding of plant biochemistry through basic research. Unlike many of the other crop improvements we've discussed, genetic engineering and synthetic biology are really the only viable approaches for these ambitious ideas.

This work is still at the stage of looking at a biochemical pathway. The journey from engineering an enzyme to increasing agricultural productivity in the field will no doubt be a very long one. Given the incredible advances in molecular biology, it now seems possible that the current bold ideas could one day become a reality.

Chapter 8

GM Animals

I keep two pet guinea pigs. They're fat, tame and multi-coloured but you can see the similarity to their wild ancestor, the montane guinea pig. The relationship between my brother-in-law's silky chickens and the red junglefowl is a bit more of a stretch of the imagination. As for my uncle's border terrier – the domestic dog's genome is so different to that of modern day wolves that we're not even sure of its origin.

For this reason, Professor Alison Van Eenennaam from the Department of Animal Science at the University of California, Davis, is sceptical about the phrase genetically modified animals: "The reason I hate that term is because I don't know what it means. I don't know what a genetically modified animal is because I think my dog is a genetically modified animal. I think if you look at what animal breeders have done to our domestic dogs you could argue that a chihuahua and a great dane are genetically modified versions of each other. It's a very amorphous term."

In the eyes of the law, however, there is a very strict definition of GM animals, and there are some under development. Scientists are already working on genetically modified cattle with resistance to diseases such as BSE and mastitis, for example, a potentially-fatal infection causing udder inflammation. Mastitis costs the US dairy industry US$2 million per year, causes pain to the animals, and leads to lower quality milk.

Alison sees a shift in the location of research into genetically engineered animals to less developed nations, which is where the greatest increase in the demand for animal protein is likely to be. Her trips to research centres in Argentina and China led her to be confident that there will be adoption of this technology, though perhaps not in the USA.

So far, just one GM food animal has gained regulatory approval: the AquAdvantage salmon.

It is thanks to modern science that meals at my grandmother's house often involve generous quantities of salmon mousse. Disease control, selective breeding for fast growing fish, nutritious feed pellets and modern technology have all made salmon farming possible, but it comes with an economic and environmental price. Salmon are predators, and much of what they eat in farms is forage fish caught from the wild, raising obvious concerns for fish stocks.

As the fastest growing food production system in the world, salmon aquaculture has become a focus for the World Wildlife Fund, which says: "The rapid expansion of salmon farming has not come without impacts. Poorly managed salmon farms can spread disease and parasites, overuse of chemicals and fish waste can affect ocean ecosystems, and the use of marine organisms for feed impacts natural resources. As demand for farmed seafood increases on a finite planet with limited resources, production systems will have to become more efficient and do more with less."

This 'more with less' sentiment is often associated with good business sense, and salmon have already been bred to grow faster. The American company AquaBounty has taken this further with genetic engineering, creating fish with the potential to grow to market size in half the time of conventional salmon.

AquAdvantage salmon are Atlantic salmon with the addition of a single gene from the Chinook salmon, which codes for a growth hormone. Some ocean pout DNA acts as a promoter, ensuring the growth hormone gene is always switched on. A wild salmon can't afford to be voraciously hungry in the winter – there's simply not enough food. In captivity, on the other hand, there's food available all year round. In this situation you want a salmon which will grow consistently, without wasting time with slower growth over the winter, and a growth hormone gene which is always switched on allows this to happen.

The recipe to create AquAdvantage salmon is simple: 1. At spawning time take female salmon bred for conventional farms (choose the prime specimens) and male salmon carrying the AquAdvantage gene. 2. Massage the fish to extrude the eggs and sperm. 3. Mix them up (you're advised to wear gloves). 4. 'Pressure shock' the fertilised eggs so the fish are sterile and unable to reproduce. 5. Store the eggs in incubators, and when the black eye dots are visible, they're ready for shipping. The result is a salmon which reaches full size in 18 months, compared to the three years of conventionally bred salmon.

Alongside the GM fish, AquaBounty has been developing a second innovation: techniques for growing fish in land-based farms. This avoids the problem of fish escaping into the wild, which is a major risk of current salmon farms which keep the fish in pens in the ocean (jellyfish attack is apparently one of the threats to these tanks…). Currently, AquAdvantage salmon are being farmed in the highlands of Panama.

The story of the AquAdvantage salmon began in the late 1980s with work at Memorial University of Newfoundland, Canada. The decades since have seen not just technological advances but also regulatory limbo, and a regulatory package costing over US$60 million. In the last few years, this seems to be paying off.

On 25th November 2013, Environment Canada approved the product for salmon egg production for commercial purposes in Canada, and in November 2015 the US FDA approved AquAdvantage salmon for sale to US consumers. Staggeringly, this approval came 20 years after the data was first submitted, and five years after an FDA advisory panel stated that the salmon is "highly unlikely to cause any significant effects on the environment" and is "as safe as food from conventional Atlantic salmon".

This doesn't mean it's all plain swimming. Not only will it take a couple of years to make fish available for the market, but the end to regulatory limbo was followed by the start of a labelling limbo. Debate continues over if and how AquAdvantage salmon will need to be labelled. There's also continuing pressure from environmental groups, which has led to some food chains announcing that they won't sell the salmon. The prospect of a genetically modified fish has caused emotions to run high, and there has even been a lawsuit filed against the FDA following AquAdvantage's approval. If a Google images search is anything to go by, plenty of myths are still circulating around the fish, such as the salmon will grow larger than conventional varieties or that they are often deformed.

Some of the vocal critics include American politicians. "I don't even know that I want to call it a fish" was the view of Alaska Senator Lisa Murkowski. During the regulatory process, over 40 members of congress called upon the FDA to halt the approval of AquAdvantage salmon. Particularly vocal was Republican Don Young who has served as the representative for Alaska since 1973. He told the Washington Post: "You keep those damn fish out of my waters. It will ruin what I think is one of the finest products in the world. If I can keep this up long enough, I can break that company, and I admit that's what I'm trying to do." It is still somewhat unclear how he thought sterile Atlantic salmon in contained tanks in Panama were going to reach Alaska and destroy their stock of Pacific salmon.

Don Young's understanding of different salmon species might be limited, but Alaska's fisheries do raise a relevant question over escapes. Alaskan fishermen have caught Atlantic salmon which have escaped from less well guarded conventional fish farms, which is concerning for local fish stocks. The Atlantic salmon haven't become widely established, but are a sign of yet another environmental challenge facing the fish industry. With less risk of escapes than from conventional farms, there could actually be reduced environmental risks should coastal salmon farms be reduced following the introduction of AquaBounty's products. Efficiency is another benefit of AquAdvantage Salmon, which grows to market size using 25% less feed than other farmed Atlantic salmon.

AquAdvantage has the potential to reduce environmental damage, but will it make salmon farming an environmentally friendly business? Not at all. AquaBounty Technology's promotional video may say "we specialise in the production of environmentally sustainable, high-quality seafood" but sustainable is pushing it. Salmon farming still has environmental impacts, however efficient your fish are.

Although AquAdvantage won the prize for the first GM animal to be approved for food, it wasn't the first GM animal in the shops. The first transgenic animal to make it to market was GloFish, released in 2003. These aquarium pets come in six colours, from 'electric green' to 'galactic purple'. Each fish has been modified to produce a fluorescent protein, and the different colour proteins come from species of coral and jellyfish. Sadly for aquarium enthusiasts elsewhere, they're only available in the US, where they can be bought in every state except California. There is no area of the US government which controls genetically engineered pets, so the company avoided the costly regulatory process.

GM animals also have great potential in medicine, and the first pharmaceutical from a genetically engineered animal was approved for use in the European Union in 2006. The drug, ATryn, is used on patients suffering from antithrombin deficiency, an inherited blood-clotting disorder. ATryn is produced in the milk of genetically modified goats. The first goats were made by injecting the human antithrombin gene into a fertilised egg, along with a section of goat DNA which makes sure the protein is produced in milk. This is an inefficient process, so the scientists moved to cloning, and now goats in the ATryn herd breed naturally.

This kind of drug is often produced using cells rather than whole animals, but a drug production facility using hamster cells can cost around to US$400 million to US$500 million, so there is an obvious desire for a cheaper option. Amazingly, one genetically modified goat can produce the same amount of antithrombin in a year as 90,000 blood donations.

Approval from the US FDA followed in 2009, though, predictably, not everyone is convinced. The Humane Society of the United States described the process used to manufacture ATryn as "a mechanistic use of animals that seems to perpetuate the notion of their being merely tools for human use rather than sentient creatures."

Given that most of us choose to take medications tested on animals, it seems unlikely that such criticisms will stop this work moving forwards. Certainly there are new developments on the way. The company which brought us ATryn has an interesting pipeline of drugs, including a project to produce a malaria vaccine antigen in the milk of goats.

Even beyond drug production, the potential for GM animals in medicine is huge. The tragic number of patients dying on the waiting list for organ transplants makes organ donation an obvious example of a challenge GM animals could tackle. The practice of patching up humans with animal parts began long before we even knew of the existence of DNA. In 1668, Dutch doctor Jacob van Meekeren performed the first bone graft on a soldier with a damaged skull, using bone from a dog. The pioneering experiment was a success, although this turned out to be unfortunate for the patient. The Church excommunicated him for being part dog, and when he asked for the graft to be removed he was told it had healed so well it was impossible to take out.

Skin grafts were also widely, if less successfully, attempted. Frogs were a popular source of skin in the 19th century. They may not be your top choice, but they had a major advantage over alternatives such as sheep, rabbits and pigeons: frogs were generally skinned alive then grafted immediately, but other donors would be strapped to the patient live for a few days!

Scientist Serge Voronoff's more imaginative experiments in the 1920s included transplanting slices of chimpanzee or baboon testicle into the testicles of older men to restore their 'zest for life'. This proved amazingly popular, and hundreds of operations were performed in the USA and Europe.

Things got a little less wacky in the late 20th century. In the 1960s primate organs were used for human transplants, and the moderate successes included a man returning to work for nine months with a chimpanzee kidney. Perhaps the best known example of xenotransplantation is the case of Baby Fae who received a baboon heart transplant in 1983. The surgery was a success but her body rejected the foreign tissue and she died 20 days later. Alongside these stories came publicity, and many people were uncomfortable with killing primates as a source of organs.

Using species of animal which are already killed for meat is perhaps a more acceptable choice. Using our more distant relatives does, however, bring potential problems such as whether the organs respond to human hormones. An even greater hurdle with using organs from other species is the ferocity of the body's immune rejection. Pig organs with lots of blood vessels tend to be rejected within 15 minutes. In fact, the last time a doctor transplanted a pig heart into a person, in India in 1996, he was arrested for murder.

An alternative line of research is to replace failing organs with mechanical components, although this comes with risks such as infection and blood clots, not to mention the technical challenges. There is also the idea of growing human organs in the lab. Exciting as the possibilities of organs grown from human stem cells may be, we're a long way from realising them.

A source of organs genetically modified to be compatible with humans would truly be a revolution for medicine. Not only would they be available on demand, they could reduce the problem of immune rejection and decrease the risk of disease transmission.

A pioneer in the field is entrepreneur Martine Rothblatt, founder of Sirius satellite radio. Satellite technology might seem like an unlikely background for a xenotransplantation expert, but she has both a personal connection to transplantation and a history of embracing blurred lines. She has spoken widely about her transition from Martin to Martine in the 1990s. She had always felt that her soul was female and now feels like she changes her gender about as often as she changes her hairstyle. She said: "People may have the genitals of a male or female, the genitals don't determine your gender or even really your sexual identity. That's just a matter of anatomy and reproductive tracts. People could choose whichever gender they want if they weren't forced by society into categories of either male or female, the way South Africa used to force people into categories of either black or white."

When Martine's daughter Jenesis found herself unable to even walk up the stairs, she was diagnosed with the rare and usually-fatal disease pulmonary arterial hypertension. Martine described her experience at the Children's National Medical Centre in Washington DC: "The head of paediatric cardiology told us that he was going to refer her to get a lung transplant, but not to hold out any hope because there are very few lungs available, especially for children."

Jenesis ended up in intensive care wards for weeks at a time. While she was sleeping Martine would visit the hospital library, starting with college text books, and she became an expert in the disease. She discovered there was a drug which aimed to halt the progression of the disease, and the drug had just been acquired by Glaxo Wellcome (now GlaxoSmithKline following a merger). Unfortunately, the company had made the decision not to develop medicines for rare diseases. Martine wasn't going to let this opportunity go by: "I went to Glaxo Wellcome, and after three times being rejected and having the door slammed in my face because they weren't going to out-license the drug to a satellite communications expert, they weren't going to send the drug out to anybody at all, and they thought I didn't have the expertise, finally I was able to persuade a small team of people to work with me and develop enough credibility. I wore down their resistance."

Despite university chemists predicting the drug wouldn't work, it was turned into a medicine which helped Jenesis who is alive and well today. Others, however, are not so lucky. The drug only slows down the progression of the disease, so thousands of patients still die each year. The only cure we currently have is a lung transplant, and Martine is now tackling the problem of the limited supply of organs.

Like most people studying xenotransplantation, her team of hundreds of scientists have chosen pigs as the donors. Even though the pig's ancestors diverged from ours around 80 million years ago, its genome is relatively similar. Its organs are also comparable to ours in both size and function, and it has the huge advantage of being easily available. The challenge is to add human genes into the pig's genome and remove some of the pig genes to make the organs more compatible with our bodies. Martine's team is making progress, and they have successfully kept pig hearts alive in baboons (we might be moving away from non-human primates as sources of organs, but at this stage they are still being used for research).

The cost of developing xenotransplantation is, unsurprisingly, colossal. Muhammad Mohiuddin, a transplant surgeon and researcher at the National Heart, Lung, and Blood Institute in Maryland, described Martine's input in the field: "She is the one that has rejuvenated the field. She has the money and a personal attachment. She wants to get it done fast."

Just as dialysis can keep patients with kidney failure alive while they await a transplant, a first step in organ transplants could be for pig livers or hearts to be used as a temporary measure, lasting months rather than being a permanent solution.

Perhaps the most likely candidate to be the first available GM animal organ is the cornea. The risk is much lower than it would be with a heart or a kidney and, with an estimated four million patients needing corneal transplants in China alone, demand is high.

The possibilities of GM animals go even further than organ transplantation, and it may well be that cells from transgenic animals are used in the clinic before whole organs. Transplanting pancreatic cells from transgenic pigs could potentially free diabetic patients from insulin injections and continuous glucose-level monitoring. It's one of the best studied areas, but trials in monkeys show this is still a work in progress.

For the millions of people who suffer from neurodegenerative disease, such as Parkinson's, cells from human embryos could be used to reduce their symptoms. However, this is generally deemed ethically unacceptable, so genetically engineered pig embryos could be an alternative. European scientists have reported some success transplanting genetically modified pig dopamine-producing cells into the brains of monkeys with symptoms similar to Parkinson's.

Probably the greatest fear with xenotransplantation is the transfer of diseases. To tackle the risk, pigs would have to be kept in strict conditions with regular disease screening. If you're in need of an organ transplant a very small risk of infection may seem a no brainer compared to the alternative, and actually the risk of infection is likely to be lower than from a human transplant. Though what about the much scarier possibility of pig diseases being transmitted to humans and into the wider population? Pig genomes are full of retroviruses – as natural genetic engineers some viruses can copy their genetic material and insert it into a host's DNA. This is a disease risk for humans, and work at Harvard University has shown that it's now possible to remove all these. This is thanks to the new technique CRISPR, which we discussed in Chapter 4. With rapid advances in molecular biology, some xenotransplantation dreams may start to seem a lot less wacky.

We don't need to be receiving transplants to be at risk of animal diseases, as some recent disease outbreaks have demonstrated. Chickens vastly outnumber humans on our planet, and one consequence of keeping so many birds in close quarters is the risk of disease. Avian influenza, or bird flu, has become an issue around the world. In 2014, outbreaks caused producers to cull nearly 600,000 poultry in South Korea, 120,000 in Japan and 6,000 in the UK, and one of the most staggering outbreaks affected 5.3 million commercial chickens in Iowa in 2015. Professor Helen Sang from the Roslin Institute in Edinburgh has dedicated part of her career to genetically modifying chickens so they are resistant to bird flu. She explained her motivation: "Not only is it a real problem to the chickens who get sick and die, and to the farmers who produce them because it causes huge losses, but it is also a potential source of human pandemic flu."

The idea is to create chickens which produce small molecules to stop the flu virus replicating. Helen's first success was to create a chicken which couldn't transmit the disease, though it still suffered from the flu. The team is now working to make birds fully resistant to infection rather than just blocking transmission between birds, so GM chickens aren't about to hit the market quite yet. If the chickens get to the regulatory stage, there will no doubt be familiar hurdles to overcome from the regulatory process itself and from those who disagree with the technology.

Peter Melchett, policy director for the UK's Soil Association, responded to the prospect of disease-resistant chickens by saying: "Keeping animals cramped together in inhumane factories encourages the spread of diseases such as bird flu and swine flu. This GM fantasy simply tries to cover up for flawed farming practice."

In reality, farming practice isn't the whole story. Bird flu has also been found in 'backyard' flocks, and migratory birds are another way the disease is spread. However, he's certainly not the only person to question the ethics of our poultry industry, and farming practices such as transporting live chickens do indeed contribute to the spread of the disease. Whilst there is good reason to call for changes to the way birds are farmed, this won't necessarily eliminate the risk of bird flu.

As far as changes through genetic modification go, Helen sees the future possibilities beyond her chickens as extremely broad. Genetic techniques could be used both to increase animal welfare and reduce economic losses. She is particularly excited by the potential of genome editing. This technique will mean that, if one breed of chicken is naturally resistant to a particular disease, the gene causing resistance can be accurately altered in commercial breeds.

She's also interested in creating a chicken which is better at digesting its food. At the moment, lots of the energy in feed is wasted because the chicken can't digest it. Helen said: "If we could add new enzymes into the chicken gut so they can break down these plant materials to release the energy, then maybe chickens will be able to eat lower quality foods and not be so expensive in terms of using high quality grains."

The domestic chicken is a prime example of both benefits and problems resulting from selective breeding. Chicken breeders have been particularly successful in creating larger chickens, and the ones we now rear for meat grow to about four times the size they did in the 1950s. They are also more efficient – you now need about half as much feed to produce the same amount of meat. However, it seems that the demand for cheap meat has been stronger than the demand for better animal welfare, and breeding has often put the chickens' skeletons and immune systems under stress. Only more recently breeders have turned their attention to selecting for more robust chickens, which should be better for welfare and also reduce the impact of disease.

Cows have also been the focus of genetic modification programmes to combat disease. Sleeping sickness (also known as African trypanosomiasis) is caused by a parasite in sub-Saharan Africa called *Trypanosoma brucei*. The first symptoms of the disease are relatively mild, including fever, headaches, joint pain and itching. The real harm is done when the parasite crosses from the blood into the brain, causing mood swings, confusion and poor coordination. In the advanced form of the disease people's sleep cycle is disrupted and they experience insomnia at night and sleepiness during the day, hence the name. Sleeping sickness is treatable, but current drugs are expensive and accompanied by serious side effects. Without the drugs, infection is often fatal. Vaccines aren't an option because the parasites change too fast.

If we put the terrible human suffering to one side, *Trypanosoma brucei* is an incredible organism. A tsetse fly ingests the parasite when feeding on the blood of humans and other animals. The parasite then travels to the fly's salivary glands. It multiplies in body fluids such as blood and lymph, and the cycle continues when the person is bitten again. This fly can then go on to bite others and pass on the disease.

One animal which is susceptible to some of the same trypanosomes as humans is the cow. The disease has made it almost impossible to raise cattle over large areas of Africa, with serious consequences for farmers. In the areas affected, about 90% of land is worked by hand.

In 2006, Professor Jayne Raper, a biochemist at New York University School of Medicine, learnt of the problem at a conference in Nairobi, Kenya, when she met scientists trying to breed resistant cattle. She explained the situation: "They're not great big cattle herds like we have here in North America. People want to keep maybe five to ten cattle so it's not as though we would instantly turn Africa into the mid-west if we could ameliorate this disease. It would just give the poorer people in Africa a means to get out of their poverty and actually use these cattle as a way of making money."

A resistant cow which didn't act as a reservoir of the disease, could ultimately reduce the number of cases of sleeping sickness in people and animals.

One breed of cattle, the hardy N'Dama from West Africa, is naturally tolerant. They're tolerant rather than resistant, so they carry the parasite, they just don't get sick. This means they still act as a reservoir for human disease. Also, the cows themselves are small and not very strong, and they don't produce good milk. Jayne explained: "Whilst they are in essence a cow that is tolerant to the disease, the farmers don't want them. They want these big, muscly, well-breeding, great milk producing cows that the Europeans brought in."

Jayne's new collaborators had breeding programmes which crossed desirable cow breeds with the resistant one. The trouble was that multiple genes are involved in the disease resistance, so it was very difficult to get all the genes to end up on one cow which also had the other favourable characteristics.

Baboons, on the other hand, are entirely resistant thanks to a single gene, coding for the trypanosome lytic factor protein. Humans produce a trypanosome lytic factor, which is from the group of proteins colloquially called 'good cholesterol'. It's pretty effective at killing many trypanosomes, just not those responsible for sleeping sickness. The baboon version is a perfect option for making a breed of cow which is fully resistant. So what Jayne needed to do was identify the baboon gene and isolate the DNA. Having done this, she created a genetically modified mouse resistant to the trypanosome which is infecting cattle.

If they were going to make use of the gene, Jayne had to work with collaborators in Nairobi. She said: "Literally the day we put the genes into the mice and we saw they were resistant, I picked up the telephone and I called them and I said 'I have the gene for you!' Would you be interested in making a transgenic cow? They were so excited because they'd been trying to do this for years."

And so their transgenic cattle project began. The team received US$2 million from the Bill & Melinda Gates Foundation and the US National Science Foundation.

Of course this doesn't solve all the challenges of raising livestock in Africa with its heavy burden of tropical disease. And it is very hard to predict the reaction to transgenic cows, which is perhaps the biggest question. Some people may also want to consider whether it is right that solutions are coming from western scientists. I personally am very pleased that money from my taxes goes to projects designed to benefit the developing world, and believe that it is appropriate for scientists in the west to use their skills elsewhere. However, there are of course plenty of examples of foreign interventions having had opposite effects to those intended, such as long-term food aid creating a cycle of dependence. But then again, western science has also brought advances in healthcare which poorer countries could never otherwise have seen. No doubt people will differ in their views on whether this intervention is a potential game changer for African farmers or a step too far.

A very different option for reducing the burden of vector-borne diseases is to reduce the populations of the insect vectors. A small company in Oxford, UK, has taken an old technique and used genetics to increase the possibilities. Fourteen years after the company was founded, Oxitec has trials on three continents aiming to reduce mosquito numbers.

Their work is based on the sterile insect technique, which was pioneered in the wake of World War 2. After radiation had been used to such devastating effect, people were now investigating more positive opportunities. This included cancer treatment and, less obviously, sterilising insects. If you release sterile insects and they mate with wild individuals, they won't produce viable offspring and so the insect population decreases.

The first successful use of the sterile insect technique was to control screwworm flies in the 1950s. The New World screwworm is a large fly whose maggots eat living flesh. The fly lays eggs on the wounds of animals such as cattle (and yes, occasionally people), munching away at living flesh. Releasing screwworms rendered sterile by irradiation proved so effective that the problem has been eliminated from the USA, and the pioneering scientists received the 1992 World Food Prize. The success of this project, combined with the drawbacks of irradiation, was the inspiration for Oxitec's founder, Luke Alphey from the University of Oxford. Chief executive Hadyn Parry said: "Radiation is quite clumsy. You can't always tell males from females so you release both and they tend to mate with each other. Also, in only a few species can you find that sweet spot where you irradiate the insects enough to sterilise them but not so much that they can't mate. Luke realised he could do this in a more sophisticated way."

Luke's method was to insert a gene into mosquitoes which causes their offspring to die before they reach adulthood. The first target was *Aedes aegypti*, a widespread mosquito which transmits devastating diseases including Zika, yellow fever and dengue fever. An advantage of working with mosquitoes is that only females bite – so you can release males without increasing transmission rates even temporarily. The genetically modified mosquitoes are reared in the lab, and the male and female pupae (a pupa is equivalent to a butterfly chrysalis) are separated based on size. The trouble with breeding sterile mosquitoes is, well, that they're sterile. However, the gene doesn't produce a toxin, it affects the activity of other genes and disrupts cell function. It's possible to keep Oxitec's mosquitoes alive by including a particular chemical in their diet, which is effectively an antidote that isn't available to mosquitoes in the field.

Often universities will license out new technologies, but for Luke this wasn't an option: this was a field in which no companies existed. Instead he founded a spin-out company from the University of Oxford.

An initial challenge was to attract investment, not an easy task given the uncertainties involved with both science and regulations. Hadyn explained: "You can't answer investors' questions such as how much will it cost and when will it get to market. You just have to say it's a great idea and a great market, will that do for now?"

With the help of their visible passion, they have attracted investors who buy into their goals and ideals. And it's easy to see why. We don't need to be reminded of the human suffering caused directly by mosquito-borne diseases, but it's worth stopping for a moment to consider the wider implications. Many of these diseases don't have vaccines, and the cost of prevention is huge, whether this is mosquito control or interventions such as bed nets. Added to this are costs of healthcare for sufferers, and the days of work they lose. So any technology which reduces this has great potential, and a genetic approach has benefits beyond increased effectiveness. Without the need for radiation devices, you reduce the cost and avoid biosecurity problems.

The ultimate goal is to reduce disease, and also to reduce pesticide use. Hadyn explained the environmental benefits of Oxitec's approach: "If you use a pesticide you hit beneficial insects too, whereas we're just hitting our target. Also, the mosquitoes don't hang around because they can't reproduce. This method is specific and doesn't stay in the environment – that's ideal for pest control."

The first few years of Oxitec's existence were spent working on the genetics and creating mosquitoes which are as strong as those in the wild. Like the pioneering team who brought the Flavr Savr tomato to market, the Oxitec team have had far more than scientific challenges to overcome. From great idea to working product is always a bumpy road, and never more so than when you are breaking new ground and entering countries which don't even have a regulatory framework to deal with your technology.

In 2009, shortly after Hadyn took over as CEO, the first out-door trials started. Hadyn explained their situation: "If you develop a drug, you have a pharmaceutical testing system out there – all the surrounding infrastructure and regulations. But there was no regulatory system for insects. We needed a system which works for safety and looks at costs and benefits."

The team spoke with national regulators, the Bill and Melinda Gates Foundation, and the World Health Organisation. They were up against pressure groups with well-organised campaigns to try and influence the regulations. Inevitably regulatory processes aren't just based on science and evidence, there is a dose of politics in there too, particularly in Europe. If dossiers were taking a long time to be processed, the team could never be sure if delays were caused by an anti-GM lobby or just by officials needing time to understand the complex subject.

Campaigns against them didn't just target politicians and regulators, there were public slurs as well. Hadyn described a time when they were hit: "We announced we were working with the Gorgas Institute in Panama, and within months there was a press release put out saying our work had been criticised in a meeting at the University of Panama. The release was designed to look as if it came from the University, but they'd just hired a room there. When I saw this I literally charged round the building saying 'who speaks Spanish' and we called all the newspapers who'd published the story to say they'd been told a load of rubbish. It was very frustrating."

One thing which those who have campaigned against the technology have successfully achieved is to get the debate further into the public consciousness. The Oxitec team sees this as very positive, and much of their work focusses on communication. It must be said that reducing insect populations by releasing more insects does require some explaining. When they go into new areas they hold town hall meetings and also go house to house to explain what they're doing. A popular accessory for this is a box full of their mosquitoes which people can put their hands in. The mosquitoes are males so won't bite, but it's fun for the team and for those brave enough to try it.

Despite this, pressure groups have criticised Oxitec for a lack communication, and this is one of the things Hadyn finds most infuriating: "For our very first trial, in Grand Cayman, we tried to get people and newspapers interested, but nobody was. We only got two newspaper articles. There was then a myth created that we hadn't communicated anything, we'd just wandered into a field and released mosquitoes."

One particularly exciting aspect of controlling mosquitoes on an island is that it offers the potential to eradicate a mosquito species altogether from that area, and that is the aim for a second project in Grand Cayman. *Aedes aegypti* is native to North Africa, so Cayman is well outside its natural range, hence the support for an invasive species eradication. Whereas using pesticides to kill mosquitoes can be damaging for the ecosystem, Oxitec anticipates bringing environmental benefits by eradicating a species which has been accidentally introduced by human transportation.

With trails ongoing, Oxitec unexpectedly found themselves with a project: Zika. The Zika virus, named after the Zika Forest in Uganda, was first discovered in 1947. However, its appearance in Brazil in 2015 brought it to the public consciousness, and a link to birth defects gradually became apparent. Its spread caused the WHO to declare Zika a Public Health Emergency of International Concern on 1st February 2016.

This captured the world's headlines, yet no vaccine is available and there is still much we don't know about the disease and its consequences. Hadyn described the situation: "It's a fever pitch debate, but actually not many practical steps are being taken. We need a response programme that looks at a number of solutions at the same time: mosquito control, surveillance, monitoring."

Zika is spread by *Aedes aegypti*, a species Oxitec already works on, so fortuitously the team was developing technology to tackle the Zika outbreak even before it occurred. It also meant they had some of the regulatory steps in place: in 2014 Brazil's National Technical Biosafety Committee had cleared Oxitec's mosquitoes to be released anywhere in the country. A further level of approval is now needed to make mosquitoes available on a commercial scale, and Oxitec now has special temporary registration. "The press have been calling for this to be accelerated," Hadyn said. "I think Zika has helped us get through the regulatory system."

This approval is an important step, and it came alongside an endorsement of the technology by the WHO. Money is still a limiting factor, but there has been progress towards release. Oxitec is waiting to sign a contract to release mosquitoes in an area covering 60,000 people, and they are starting to put up their first major factory. As Hadyn explained, it's a long process: "You can scale the biology very quickly, because each female can produce 250 other females per month. What takes more time is hiring people, training people and building factories."

Ultimately, Oxitec intends to design efficient factories so they can produce enough mosquitoes to protect millions of people. This comes with practical questions such as how big will the cages be, how will the eggs be handled, how do you move mosquitoes? As they gain experiences of scaling up, the team constantly see ways to improve things. Hadyn described the evolving technology: "Our very first release system was someone sitting on the back of a truck and releasing mosquitoes from pots. Now we've designed a system so that pupae will go into a vessel, and there will be no more handling until the adults are released into the environment. That's our future plan."

Even once mosquitoes have been released on a large scale, the challenge will be to study how much they have reduced the burden of disease. This is much harder than it initially seems – people move around so will probably spend time exposed to the disease in areas where Oxitec's mosquitoes haven't been released. Unless release is over a very wide area, it will be hard to discover what the impact is on local transmission rates. Modelled predictions indicate that Oxitec's mosquitoes can bring disease down to a level where the disease is no longer transmitted, so Hadyn is confident. History also gives reason for hope. Campaigns to control mosquitoes with DDT in the 1950s and 1960s meant malaria was almost eliminated in Brazil, proving the potential for vector control.

Looking at the progress he has seen is his time as CEO, Hadyn said: "When you look back, you realise there has been enormous progress in a short space of time. When you look forward you see the walls that need to be knocked down."

Funding, regulations and criticism are all challenges that the team have weathered, and the future looks positive for Oxitec. With large scale trials showing signs of success, there is a sense of momentum. Recent good news includes an assessment of 'no significant impact' by the US Food and Drug Administration, which means they don't have concerns about environmental damage.

In 2015, Oxitec was acquired by an American synthetic biology company, Intrexon, and its future plans cover agriculture as well as health. Olive flies, the sole major pest of olive trees in southern Europe, are high on the list. Another target which Hadyn would love to tackle is the malaria carrying *Anopheles* mosquitoes. Sadly, this doesn't look imminent, simply because there isn't a commercial market. Only with philanthropic funding could this become a reality any time soon.

Beyond disease, another focus of animal genetic modification is environmental impact. In the 1990s, scientists at the University of Guelph in Ontario, Canada, believed they were on the way to reducing the environmental impacts of pig farming. They had created the Enviropig, a line of Yorkshire pigs which had been genetically modified so they could digest plant phosphorus more efficiently. Like us, pigs need phosphorus to make DNA, build cell membranes and transport energy, yet their diets don't naturally provide enough.

Plants store phosphorous in the molecule phytate, which must be broken down to release the phosphorous. Pigs don't naturally have an enzyme to do this; they rely on bacteria in their gut. The process is inefficient and so lots of the phytate simply passes through the pig's gut and ends up in manure. It then becomes an environmental pollutant in rivers and streams, fuelling excessive algal growth. Farmers currently supplement feed with phosphorous or commercial forms of the enzyme produced by microbes, and trials have taken place of plants genetically modified to produce the enzyme themselves. The Enviropig team took this one step further: pigs were modified with a bacterial gene to produce the enzyme in their saliva.

Years of research demonstrated that the gene was passed stably down between generations. It was also important to show that the enzyme was produced only in the salivary glands, with no more than trace amounts found in any of the animals' other tissues (and hence in the pork). Tests showed that the protein composition of the pork appeared normal so everything was looking good.

Enviropig was submitted for regulatory review in the USA in 2007 and in Canada two years later. The initial assessments seemed positive, but the reviews were lasting for years with no clear end in sight. At the same time there were, of course, pressure groups trying to slow the regulatory process, and the Canadian Biotechnology Action Network coordinated the major Stop Enviropig campaign.

In 2011, the project lost the support of their long-term funder Ontario Pork Producers Marketing Board. Unable to find alternative income, the researchers couldn't afford the costs of raising the pigs and dealing with the regulatory process.

The following May, all remaining Enviropigs were slaughtered at the University of Guelph. A local farm sanctuary offered to care for them, but the university decided this wasn't appropriate for these transgenic animals. For the scientists involved, it was a disappointing end to a promising project. "These pigs were healthy pigs that did perform as they were designed to perform. They just didn't meet the social requirement," lamented Professor Cecil Forsberg.

Many campaign groups celebrated the 'death of the Enviropig', though semen has been saved should the time ever be right for resurrection.

Chapter 9

'Superfoods' and Enhanced Nutrition

In our 10,000 year history of altering plant genetics to suit the needs of agriculture, nutritional quality is a very recent consideration. Early on we had great success at reducing the harmful compounds found in some plants, but the concept of nutritional value is a fairly modern one. Agriculture has instead focussed on higher yields at a lower cost, and there is even evidence that we have reduced the levels of essential minerals in some modern varieties of cereals and vegetables. Cereals have often displaced nutritionally superior legumes and pulses, increasing the quantity of food we can grow but often at the detriment of a varied diet. We often compound the problem by removing parts of the plant which are rich in vitamins, minerals and fibre yet have a texture many people dislike.

The consumer hasn't been completely forgotten though. After all, it's thanks to plant breeding that strawberries are so big and juicy and sweetcorn is so sweet. Perhaps the most remarkable use of plant breeding to please the consumer is cannabis. Modern varieties of 'skunk' have been bred for strength, and have high levels of the psychoactive compound THC. Ironically, breeding for increased THC has left plants producing less of the anti-psychotic compound CBD. Scientists researching medical marijuana are now working to produce plants with the benefit of high CBD.

The consumer benefits which will be the focus of this chapter, however, are nutritional. In addition to the billion people worldwide who are undernourished due to a lack of food, over two billion people suffer from 'hidden hunger'. A shortage of essential vitamins and minerals can lead to disease, impaired child development, and even death. Poor nutrition is a major global health problem, contributing to half of the nearly 10 million deaths that occur each year in children younger than 5. The problem is worst in the developing world, but developed countries have their own needs for better nutrition.

Opponents of GM are quick to point out that these problems have alternative solutions, and this is certainly true. In the developing world, supplementation programmes costing billions of dollars are saving lives and reducing suffering. Fortification of foods can also be successful, such as adding iodine to salt or vitamin A to bread. Encouraging farmers in low income countries to diversify the crops they produce can have huge health benefits.

Despite these efforts, the fact remains that deficiencies of iron, zinc and vitamin A are among the ten leading causes of illness and disease in low-income countries. Current strategies to tackle this are often hampered by lack of infrastructure, purchasing power or access to healthcare. A varied diet is undeniably the best way to tackle nutrient deficiencies, but for many people on low incomes this simply isn't an option.

An alternative approach is biofortification. This involves altering plants to increase the vitamins or minerals we gain from eating them, and is by no means limited to genetic modification. Conventional breeding can also be successful at biofortifying crops. Genetic modification expands the possibilities, allowing fortifications which couldn't be achieved through conventional breeding and creating crops which are enriched in multiple nutrients.

By far the most famous GM food designed to tackle a nutrient deficiency is golden rice. As a teenager in the late 1990s, my school textbook held it as an example of the promise of genetic modification, counterbalancing the scare stories I was hearing in the news. There are good reasons why biofortified crops are more palatable to consumers than other forms of genetic modification. For a start, we see the benefits, and biofortified crops don't come with the environmental questions associated with resistance to pests and herbicides. However, even golden rice has been hampered by protests, patents and technical challenges, and it is yet to be grown commercially. Before we explore the troubled history of golden rice, we will look at some lesser-known examples of nutritionally enhanced crops.

The root vegetable cassava is the major source of calories for over 700 million people who depend on cassava for their daily calories, most of them in South America, Asia and Africa. The problem is that a typical diet based on cassava provides less than 30% of the minimum daily requirement of protein and only 10-20% of a person's iron, zinc, vitamin A, and vitamin E requirement.

Biofortified cassava designed to tackle vitamin A deficiency was released in Africa without the world's attention, and the project continues. The crop development, through conventional breeding, was thanks to the dedication of Dr Hernán Ceballos from the International Center for Tropical Agriculture in Columbia, along with his colleagues and collaborators. They have named their crop yellow cassava (yellow rather than golden – does that say something about GM branding?) and the story provides a valuable comparison to fortification using biotechnology.

For anyone who opposes golden rice, their reasons for opposition will affect whether the learnings from this story are relevant. But for those sceptical about biofortification, you won't find a stronger advocate than Hernán. He told me his motivations: "As we learn more about the tragedy of malnutrition we've really realised that the argument in favour of biofortified crops would be improving the public health of millions of people. I realised that half a million children turn blind in Africa because of lack of vitamin A, and these children are eating cassava every day. Every time I say this, I feel embarrassed. It is a huge tragedy, and the shame is that it's a preventable tragedy."

Hernán began his scientific career in his home country, Argentina, and arrived in Columbia as a maize breeder in 1989. Ten years later he was converted from maize breeder to cassava breeder, and was disappointed that the transition came with a project on micronutrients. At the time he felt this was stupid: "I was coming with the inertia of the green revolution, that we had to produce enough food. Quantity was the name of the game, not quality. But then, and this is the wonderful thing about this experience, it exposed me to the other side of the world, allowed me to talk to nutritionists, exposed me to the figures of vitamin A deficiency, and then I saw the light. I realised the huge problem and that we had a great opportunity to make a contribution."

In Europe and North America we're most familiar with cassava starch in the form of tapioca, including the 'pearls' in bubble tea. Elsewhere, cassava is a major part of many diets, and annual global production has increased in recent years. The trouble is that cassava is often eaten as a staple rather than as part of a balanced diet. The roots are a good source of energy while the leaves provide other nutrients, but some key nutrients are missed out.

Initially, Hernán's work tackling this with nutritionists and public health experts revealed that the topic has unavoidable conflicts of interest, as passionate people came together with the belief that their work was the way to solve the problem. People already fighting micronutrient deficiency were faced with the important question of why should they be distracted when supplementation was working? Hernán and his colleagues continued to explain their justifications for tackling the problem at its source, and to reassure people that this work didn't belittle what was already happening. Eventually, some key people in the nutrition community were convinced, and the resistance was gradually erased.

Scientifically, the challenge was to develop cassava with high concentrations of beta-carotene, which the body uses to make vitamin A. Modern developments in the lab, along with expertise from a collaboration with French scientists at CGIAR in Montpellier, led to faster progress. Hernán said: "It has been a very rewarding kind of work. We improved a lot the way we do the breeding – we refined the breeding strategy to make faster progress. We used good old fashioned technologies, taking advantage of some new tools. Nothing fancy, just hard work and common sense."

For Hernán, a 'magic' new technique was near-infrared spectroscopy. Not only did it show the beta-carotene concentration in a sample of cassava, it gave the level of cyanide, which is naturally present in cassava. Cyanide is toxic, and so throughout the project they had to be sure that it would only be present at safe levels. Near-infrared spectroscopy also increased their speed, much to Hernán's delight: "Rather than having eight samples a day, we can have 100 samples a day, and this is sweet news for us."

Not all the improvements the team made were quite so carefully planned. Cassava roots spoil very quickly because the only biological reason for a root is a reserve organ for the plant. When the root is no longer attached to the plant it breaks down in just a few days. Hernán described how they discovered an unexpected benefit: "Science is also more often than not based on accidents. We were preparing a shipment of roots for bioavailability studies in the US, but there were three roots left behind. A month later the assistant went back and he carefully picked them up, thinking they would be rotten, fermented, that they would disintegrate. And, voila, the roots were perfectly fine. The antioxidant properties of the carotene slow down the deterioration."

For many years they worked to increase the beta-carotene content, which they did successfully, but that's not the whole story. It had to be followed by complex breeding work to ensure the cassava tastes good and cooks well. And from a nutritional perspective, it wasn't enough to produce a cassava with high carotene content, they had to show that this doesn't get quickly broken down, and that once the cassava is eaten the body can turn the carotene into vitamin A. Hernán's team did some retention studies, and other studies demonstrated that the carotene in cassava roots is very efficiently transformed into vitamin A.

These results were very promising, but it is much more challenging to find out what effect it has on people's health in the real world. This will depend on how much yellow cassava people eat, and on their health – diarrhoea, for example, can affect digestion. Hernán described the challenges: "These studies are really very expensive, and you have ethical problems. To do this you have to measure vitamin A levels in people, and if you find that they are deficient you have to provide a supplement. But these are the ideal people to test what will happen with the cassava."

It's certainly too early to judge the impact of the project, and the team intends to reach far more people once new varieties are available. As Hernán points out, this is one step in the right direction which needs to be accompanied by health messages such as eating more greens, and crops fortified with other nutrients may also have a role to play.

Their current work on yellow cassava aims to increase the number of varieties which have high beta-carotene content combined with good agronomic qualities, such as resistance to cassava mosaic virus. Despite the years of research which led to the development of yellow cassava and the work still to do, it seems they have beaten golden rice to generate the first benefits.

Problems which remain include iron and zinc deficiencies which both prevent healthy development in children. Iron is important for red blood cells to transport oxygen, so feeling out of breath and dizzy are common symptoms of iron deficiency (anaemia). Symptoms of zinc deficiency may include loss of appetite, diarrhoea and impaired immune function.

These impacts on people's health means Hernán has set his sights on iron and zinc fortification, and he believes that genetic engineering will be the right approach: "There is very interesting work using genetic transformation that looks very promising, and I don't have genetic variability to use conventional breeding approaches. I know people in Europe are very reluctant to think about genetic transformation, but in Europe you might have cereals in the morning that are fortified with iron and zinc so you don't have any problems of deficiency. But in Africa, particularly women are chronically anaemic mainly because of lack of iron."

Genetic engineering may be the right approach for zinc and iron, but that doesn't necessarily mean it is for beta-carotene. Attempts at increasing beta-carotene in cassava using genetic modification have proved challenging. One side effect of attempts to increase the carotene levels in cassava using genetic transformation has been to reduce the dry matter content to unacceptably low levels. Hernán believes this may be possible to overcome, but having so successfully produced carotene by conventional methods he doesn't see much effort going into this in the future.

He is, however, happy that lots of approaches are being tried: "In a way the transgenic approach is a threat for the work I do because people can go and say 'give your money to me, I will give you a transgenic cassava' and I say 'don't give money to transgenic people, give it to me I'll do the breeding work'. It doesn't really happen that way. I think that in general the people working on cassava are a small community and there is room for everybody. In a wise way, we attempt different approaches and the most efficient approach prevails."

Elsewhere in the world, there are multiple projects with the goal of improving cassava's disease resistance and nutritional content. Professor Hervé Vanderschuren from the University of Liège in Belgium is using genetic engineering to increase cassava's vitamin B6 content. Hervé and his team inserted two genes from Arabidopsis, a well-studied weed which is widely used in the lab. He said: "What is surprising, what is nice, in this story is that using the Arabidopsis gene you can reach such a high increase, 30 to 40 fold, in vitamin B6 in leaves and in roots. So it seems that those enzymes can bring this benefit in many different crops."

The experimental varieties look promising – they have elevated levels of vitamin B6, and research has shown that this can be absorbed by humans. The next stage is to introduce the genes into varieties which will suit local farmers. For this, local scientists will need to be heavily involved, and Hervé has strong collaborations in Africa and Asia. "I don't think European labs should be doing this work," he said. "I think people locally should have the freedom to select the improved traits they think are relevant for their agrosystems and fully participate in the implementation of those solutions."

To make this possible, much of his time is dedicated to training local scientists and disseminating the cassava technologies. The technologies themselves had to be improved so they could work effectively even in the conditions commonly found in African labs, where neither the lab supplies nor the electricity are completely reliable. His goal is to allow African biotechnologists to develop their own solutions for crop improvement.

Iron and zinc deficiency are unusual in that it is not just that we consume too little, but that an 'anti-nutrient' makes them less available for absorption by our bodies. This anti-nutrient, phytate, is naturally produced by the cassava plant. It is possible to reduce phytate production through conventional breeding, by using chemicals to induce genetic mutations. Genetic modification adds new options for biofortification. Enzymes can be added to break down phytate, or the plant can be modified to store more iron in its grain.

Progress has already been made with iron and zinc biofortification: high-iron and high-zinc rice has made it to the product development stage at the International Rice Research Institute (IRRI) in the Philippines. The rice varieties are predicted to be ready for release to farmers in about 2022 in Bangladesh.

It's not just humans who can benefit from biofortified crops, though; livestock are also in need of extra nutrients. Up there with the best known genetic modifications is an altered ability to produce lysine, most famously in genetically-modified dinosaurs.

The inhabitants of Michael Crichton's Jurassic Park had been genetically altered so they couldn't produce lysine, making them reliant on supplements from the vet. This was a safety mechanism to prevent them from surviving outside the park, and we all know how that went down. In reality, no animals can make lysine. Livestock and people get lysine from their diets so real life genetic engineering has turned a poor source of lysine, cereal crops, into a good source.

'Mavera maize' produces high levels of lysine thanks to the addition of a bacterial gene. Traditionally we have solved the problem by feeding soybean to livestock and adding lysine to animal feed as a supplement. This means that hundreds of thousands of tonnes of lysine are produced as feed additives each year, which in 2009 had a market value of over €1.22 billion. Currently this lysine is produced by bacteria, which may themselves be genetically modified. It is a gene from the same bacteria which has given Mavera the ability to produce lysine, and the effect is a maize variety which is not only cheaper but also a more nutritious animal feed, reducing the need for supplements.

Rothamsted Research's Professor Nigel Halford thinks this should sound alarm bells for European farmers who may find themselves competing with crops like Mavera but unable to grow them because of EU regulations. He also believes this should be food for thought for anyone who is keen to make the EU less reliant on imports. He said: "Soybean has displaced barley in livestock feed because of its lysine content. A high lysine barley could make the EU more self-sufficient in animal feed, but the current climate in the EU means nobody would put money into varieties suitable to be grown here."

Maize has also been modified to enhance growth in livestock whilst reducing phosphorus pollution. Livestock can't digest the form of phosphate normally found in plants, so they excrete it instead, causing phosphate pollution. Scientists have modified maize to increase the production of an enzyme which helps the animals digest phosphorus, and this became the first genetically modified maize officially issued with a biosafety certificate in China.

Closer to home, biofortification is tackling not nutrient deficiencies but diseases associated with over-consumption. Scientists from the John Innes Centre, a research institute in Norwich, England, have developed 'purple tomatoes'. The dark pigment, anthocyanin, is intended to give tomatoes the same potential health benefits as fruit such as blueberries. Anthocyanin is an antioxidant which studies on animals show could help fight cancer, and it is produced in purple tomatoes thanks to a gene from a snapdragon plant.

Large-scale production of the purple tomato is now underway in Canada where the regulatory process is faster than in Europe. Tomato juice will then be imported to the UK for scientists to test whether it has the health benefits hoped for. Volunteers will be given twice weekly portions of purple tomato soup and their health will be monitored. They won't know whether their soup is purple because of the tomatoes used or because of food colouring.

The purple tomato team is now experimenting with other antioxidants in tomatoes, and is also looking to increase the health benefits of the orange. The beautiful red colour of blood oranges likewise comes mostly from anthocyanin pigments. What particularly sparked the scientists' interest in blood oranges was studies of mice fed on high-fat diets. Blood orange juice limited their weight gain, enhanced their sensitivity to insulin and decreased their cholesterol levels.

Oranges are thought to have originated in Asia as a cross between the pomelo and the mandarin. A mutation to these gave rise to blood oranges, with their darker colour. We don't know much about when or where this happened, but the first written record of a blood orange is in 1946 from Giovanni Baptista Ferrari, a member of the Jesuit Order in Rome.

The beneficial anthocyanins are at their highest levels in the dark red 'Moro' oranges, but making these benefits widely available comes with major challenges. For a start, many people find the Moro bitter and hard to peel. It's also impossible to provide large quantities. Although blood oranges are grown around the world, those grown in Sicily have particular health benefits. Sicily's unique climate, particularly the cold nights, causes the oranges to produce more of the beneficial pigments. This led the scientists to wonder whether they could make blood oranges which were as palatable to the consumer as sweet oranges and have the same health benefits wherever they are grown.

The team, led by Professor Cathie Martin, looked at all the types of blood orange known from around the world to see if any produce anthocyanins without being exposed to the cold. However, their results offered little hope of using conventional breeding to create blood oranges that are free from cold dependency.

In blood oranges, the anthocyanin genes are only switched on at low temperatures. The region of DNA which acts as a switch to the gene is sensitive to temperature. To ensure this gene is active whatever temperature the orange is grown at, the GM variety has a switch which isn't temperature sensitive.

This work is still in an early stage, and we're yet to see whether the quantity of juice required to gain sufficient amounts of anthocyanin is reasonable for people to drink. Our varied and complex lifestyles also mean that demonstrating the benefits to humans will be much harder than giving controlled diets to genetically similar mice. Some people are also likely to question the whole premise. The project is designed to tackle health problems associated with excess fat-consumption, so arguably reducing fat intake is a better way forward. With obesity at the level of a public health crisis, however, the oranges remain an appealing option.

Other biofortification projects are tackling environmental as well as health challenges, and an ingenious plan was hatched following conversations between crop scientists and fish nutritionists working with the aquaculture industry. A team at Rothamsted Research, led by Professor Johnathan Napier, has spent nearly two decades working on a project to meet the needs of the fish farm industry.

We are all aware of the health advice to eat oily fish, and the benefits come partly from long-chain omega-3 polyunsaturated fatty acids. These help prevent heart disease and stroke, and there is even evidence that they're linked to mental health. Sadly, the advice to eat more fish can be in conflict with environmental sustainability.

The collapse of fish stocks and the plight of the marine environment is perhaps the most widely understood environmental damage caused by our love of animal protein, yet we're a long way from solving the problem. I spoke to Johnathan on the same day as an FAO State of World Fisheries and Aquaculture report was released, with the conclusion that more should be done to reduce overfishing. The report showed an increase in fish consumption, whilst there was no improvement in the state of the world's marine resources.

Over half of the fish we consume comes from farms, and much of the fish caught from the wild goes to feeding them. Aquaculture uses around 80% of the fish oil harvested annually from the sea, and more fish goes into the system than comes out. We use wild fish as feed because, ironically, fish can't make fish oils. The oils are in fact made by microalgae, which aren't available to caged fish in farms. We don't currently have the technology to produce algae on a large scale, so farmed fish are fed oil and meal made from their wild (usually smaller) cousins – analogous to what happens in the oceans. It is mostly the smallest fish which eat the omega-3-rich algae, and these fatty acids accumulate all the way up through the marine food web. Although it's perfectly possible to raise fish with vegetable oil in their diets, this changes their nutritional value. Fish can only accumulate fish oils if that is exactly what they've been fed. As Johnathan said: "Fish are a classic example of you are what you eat. Now that farmed fish have more vegetable oil in their diets, consumers don't get the healthy meal they expect."

In the search for a more sustainable source of fish oil, Johnathan has used genetic modification to produce 'fish oils' in plants. His long-term aim is to provide oils for farmed fish not from the sea but from our fields.

The project uses Camelina sativa (false flax), which is a distant relative of oilseed rape. Camelina is naturally high in short-chain omega-3, though it's the long-chain omega-3 which has the health benefits. To change the profile of oils, the team has introduced synthetic DNA sequences into the plant, similar to the genes found in algae. These new genes produce enzymes for the biochemical pathway which makes the oils we're after. The long-chain omega-3 oil can be extracted from the seed and fed to fish in a pure form.

In the laboratory, hundreds of camelina plants are producing the long-chain omega-3 oils in their seeds. Small outdoor trials are now taking place to see whether the same is true in the field. Camelina is predominantly self-pollinated but is visited by insects, so Rothamsted made efforts to address concerns from local beekeepers by covering the flowering crop with a fine mesh net to prevent bees from transporting GM pollen back to their hives. Such measures wouldn't be feasible if the crop was grown on a commercial scale, but the team is now starting to study the farm management practices which would be needed to prevent gene flow. Thankfully, camelina doesn't cross pollinate with oil seed rape or other Brassica species, making large-scale planting more feasible without risking cross-pollination of other crops.

The results from trials so far look very promising: it seems that the plants perform as well in the field as they do in the greenhouse. Salmon feeding trials using oil produced from greenhouse-grown plants have also seen positive results, with fish accumulating fatty acids exactly as they would when fed fish oil. Johnathan is now looking at the next steps towards the idea becoming a commercial reality: "We know that oil from our camelina works as a diet for fish, and we're now tailoring it to be even more suitable for feed. In the future we also hope to do bigger trials in multiple locations to determine how well the crop grows in an agricultural environment."

In theory, if genetic modification allows us to replace animal sources of nutrients with plants, this could be good news not only for animal welfare but also for the environment. It would even be possible to skip the step of aquaculture altogether by using the plants to create supplements for human consumption, although this isn't the path currently being followed as health professionals recommend consuming omega-3 as part of a whole food. Growing fish oil on land could reduce the tremendous pressure the fishing industry puts on the oceans, whilst providing a scalable supply to better match the needs of aquaculture. Still, there is a long way to go before the technology becomes commercially available. For Johnathan and his colleagues, however, the current trial is a major breakthrough. He said: "We're another small step towards a commercial crop in the field. I'm not expecting this to be an easy solution overnight. But even if it reduced the strain on oceanic fish oils by 10-15%, that would be a good thing."

A key difference between Johnathan's work and many GM biofortification projects is that there is a commercial market for fish oil, which is in sharp contrast to the original use of genetic engineering for nutritional enhancement. All the products and experiments we have considered so far come in the wake of GM's most famous invention: golden rice. Named due to the rich yellow of the rice grains, golden rice has been developed as a humanitarian project to reduce vitamin A deficiency.

Vitamin A deficiency is considered a health problem in over half of the world's countries, causing disruption to the immune system and increasing the severity of childhood diseases such as measles. Vitamin A is essential for the retina, so deficiency can lead to blurred vision and eventually blindness. Vitamin A is naturally found in leafy vegetables, and a major cause of deficiency is over-reliance on white rice. But what if this rice was golden?

In 1999, two professors announced that they had created golden rice, which they believe holds the potential to save the sight and lives of over a million children. Ingo Potrykus and Peter Beyer had fortified rice with beta-carotene, the precursor of vitamin A.

The plan to tackle vitamin A deficiency with biofortified rice was hatched in the 1980s, but genetic modification of cereal crops initially proved harder than expected. Even the 1999 strain was still a prototype and wasn't high enough in beta-carotene to be useful as a product. In the following years the first golden rice had to be modified to produce higher beta-carotene concentrations. Originally, the two enzymes needed for the plant to produce beta-carotene came from the daffodil and a type of bacteria. Daffodils were chosen because the enzyme is present in high concentration, but as laboratory techniques improved the enzyme from the daffodil was replaced with a more efficient one from maize.

Early scientific difficulties were only part of the problem. In the first decade of the 21st century, Potrykus and Beyer faced unexpected legal challenges. Their intention had been to patent golden rice and allow it to be used for free. However, they had used techniques patented by other people so their finished product would be bound by these. There were a staggering 70 patents, though thankfully the large corporations which owned the patents quickly granted the rights to offer the technology free to poorer farmers. Anyone earning less than US$10,000 a year can replant seeds and trade them within their community.

An ongoing challenge is that approvals for large-scale release are extremely time consuming. The regulatory process is lengthy for all GM crops, particularly in countries where adequate regulations are only now being put in place.

Progress has also been slowed by opposition groups. As recently as August 2013, field trials were attacked by activists. In this case they were young Filipino city dwellers who said they represented farmers. Sadly, allegations are being investigated that they were paid by opposition groups.

Dr Robert Zeigler, director general of the International Rice Research Institute, responded by saying: "Activists raise concerns that we are all interested in, which science can help address, but if such scientific research is shut down, then we cannot answer the important questions on the safety and huge potential benefits of golden rice. We should all support the science-based assessment of golden rice and allow national regulators to undertake their important work without international pressure or interference."

Activism against golden rice hit the headlines in 2016 when 110 Nobel Laureates signed an open letter to Greenpeace calling on them to abandon their campaign against GMOs, and against golden rice in particular. Given that there are fewer than 300 living Laureates, literature included, this caused quite a splash. GMWatch and other allies responded by referencing social anthropologists who had concluded that GMO opponents weren't the problem here. This position has been hotly disputed, though in reality it is impossible to know how much damage activists have done to the case of golden rice. Did they scare away research dollars? Are they responsible for regulators setting extremely high hurdles?

Golden rice is still under evaluation. Although the research is yet to be completed, it appears that golden rice has the potential to reduce poverty, disease and malnutrition. There are, however, groups who consider alternative solutions to be more effective.

Greenpeace and other environmental groups promote homestead food production, along with supplementation, as an alternative to golden rice. Creating balanced diets through small-scale growing of different vegetables certainly has the advantage of providing other nutrients too. However, it is not currently feasible as an alternative to supplementation or golden rice. Advocates of golden rice would see home farming as complimentary; the Bill and Melinda Gates Foundation invests both in golden rice and in homestead food production.

Connected to this is a belief that we should address the underlying reasons for vitamin A deficiency rather than just treat the symptoms. Poverty is both a cause and an effect of vitamin A deficiency, which complicates this argument.

One of the most common criticisms is cost, and exactly how much golden rice will cost per person reached is difficult to predict. So far the great expense has yielded no return, but if uptake is high then the cost savings will be considerable. Farmers will be free to cultivate the rice and pass on their seeds, removing the reliance on regular deliveries of vitamin A supplements. Research predicts that biofortified crops can be a cost-effective solution.

Supplementation programmes are a very expensive business, and currently vitamin A supplementation costs over US$500 million each year. The cost of vitamin A pills is tiny, but supplementation costs at least US$1.00 per person when distribution is taken into account. This figure is dependent on sharing distribution costs with vaccinations such as polio, and the success of vaccination programmes means they may wind down, increasing the cost of vitamin supplementation.

In an ideal world, the technology would become redundant as diverse diets provided all the vitamin A each person needs. Until that time, there will be an ongoing cost involved in supplementation, and golden rice has the potential to be a more sustainable alternative. If golden rice reaches more people than supplementation programmes can, their improved health will bring economic as well as humanitarian benefits.

Perhaps an even more important factor in this debate is that while researchers believe that golden rice could save millions of lives, opponents still maintain that it will be ineffective. Thankfully, we can test the assertions of both sides, and for some of the claims we already have. It is perhaps ironic that campaign groups opposing golden rice as an 'unproven technology' are acting to hamper collection of evidence to prove, or refute, its value.

Over ten years after the first golden rice was created, evidence was looking good for the claim that golden rice was an effective way to deliver vitamin A. Scientists from Tufts University in Boston led a trial in China on the effects of eating golden rice. Their study, published in 2012, showed that the beta-carotene in golden rice was as effective as beta-carotene in oil and better than that in spinach at providing vitamin A to children. A bowl of golden rice (50g dry weight) could provide approximately 60% of the Chinese Recommended Nutrient Intake of vitamin A for 6 to 8 year old children. However, this was delivered a considerable blow in 2015 when the paper was retracted.

It's not that science that was in question, it is the way the ethical procedures were followed (or rather, weren't followed). Claims by Greenpeace that the children had been used as guinea pigs prompted Tufts University to begin inquiries into the work. These unearthed concerns with the informed consent process. In particular, Tufts concluded that there was inadequate explanation to the participants' parents of the fact that golden rice is genetically modified.

The study's author, Dr Guangwen Tang, took the journal to court in an attempt to prevent the retraction. Her claims that the retraction amounted to defamation weren't upheld, and a Massachusetts judge cleared the way for a retraction.

This leaves us with the memory of a paper with an uncomfortable mixed message for golden rice. Scientifically, things are good: the university enquiries found no problems with the accuracy of the results, and there were no fears that the children in the study had been put at risk. Socially, however, it has created a black cloud. Only time will tell how this contributes to the golden rice storm.

When the time comes for release, golden rice will only be made widely available if it is approved by national regulators and shown to reduce vitamin A deficiency for the world's poorest populations. Helen Keller International (HKI) - an international NGO dedicated to preventing blindness and reduce malnutrition for the world's most vulnerable and disadvantaged - plans to independently evaluate the efficacy of golden rice if it is approved by national regulators and deemed safe for human consumption. Safety assessments from field trials are expected to be published soon.

Despite the golden rice disappointments, researchers have set their sights on enhancing other foods with beta-carotene, and of creating varieties of golden rice which are biofortified with multiple nutrients.

Professor Beyer, one of the instigators of the golden rice project, is leading the ProVitaMinRice Consortium, a collaboration between researchers in Australia, China, Germany, Philippines, Hong Kong, USA and Vietnam. The aim is to develop new varieties of rice with increased levels of pro-vitamin A, vitamin E, iron, and zinc, or with nutrients that are easier for the body to absorb. They are also increasing protein quality and content. The project plans to incorporate the new rice lines into breeding and seed delivery programs, and to make them freely available to low-income farmers in the developing world.

On a similar note, a team of scientists in Australia, Uganda and the USA are attempting to genetically modify bananas in Uganda so that their content of vitamin A, vitamin E, and iron is equal to or exceeds the required daily allowance.

Overall, these case studies have demonstrated the potential that genetic modification has to improve the nutritional value of foods we already consume. Biofortified crops come without some of the environmental and corporate issues associated with crops developed for agricultural reasons (and often with profit as a major motivation). There is no concern, for example, about increased use of herbicides or about insects becoming resistant to pesticides. As we have seen, however, biofortified crops still come with their own challenges, and there are overarching questions which some people are still asking.

For example, research into biofortified crops is often a collaboration between scientists from the developed world and those from the countries set to benefit. Are scientists from richer countries providing essential skills and funding, or should such projects be led entirely by scientists from the regions which are seeking to benefit?

There are also wider issues of whether people are in a position to benefit from the crops. I discussed this with agricultural economist Sharada Keats, who works for the Overseas Development Institute, a UK think tank. Although she sees a potential role for nutrient enhanced crops, she sees them as acting at the margins. A major cause of micronutrient deficiency is actually health problems which prevent people from absorbing or using the nutrients they've consumed. As Sharada said, "if a kid isn't keeping any food in, it doesn't matter what they're eating." Water, sanitation and access to good healthcare are needed to ensure that people are healthy enough to absorb the nutrients.

Opponents also fear the success of biofortified crops will be a 'Trojan Horse' technology, paving the way for other GM crops. Fearing the benefits of a crop seems to me one of the weakest arguments against it, but it holds weight with many people.

Ultimately, my dream of obtaining all essential nutrients from cake is a long way off, but biofortification has interesting potential for improving health in the developing world and beyond.

Chapter 10

More Than Just Food

Our history has been shaped by the use of plants for shelter, transport, fuel, medicine and warmth. We've moved beyond wooden canoes to powered vehicles, and plants have become an interesting, if controversial, option for providing this power. Cars are greedy for energy, and so the vast majority rely on liquid fuels. These are proving very hard to produce sustainably, and there's certainly a good reason for the stream of bad press about biofuels made from crops such as maize, oil palm and sugar cane. Billions of litres of bioethanol are used each year, yet there are suggestions that it can actually produce more carbon emissions than the fossil fuels it's replacing. In Brazil, which joins the USA as the world's top biofuel producers, growing crops for fuel is leading to large-scale deforestation. This is bad news both for biodiversity and for climate change.

Elsewhere in the world people rely much less on biofuels, though this could be changing. An ambitious vision for 2030 is that up to a quarter of the EU's transport fuel will come from clean and CO_2-efficient biofuels. It's a target we are an exceptionally long way from meeting, and some creative solutions will be needed if we're going to get anywhere close. Part of that creativity could come from genetic engineering.

In 2011, the US Department of Agriculture approved the corn Enogen from Syngenta, the first GM cereal specifically designed for biofuel production. The corn produces alpha-amylase, the enzyme that breaks down starch into sugars, which is otherwise added separately in the processing plant. Amylase is a widespread enzyme – we produce it in our saliva – but this gene has been taken from heat-tolerant bacteria, which means it can work at high temperatures in the ethanol manufacturing plant. This makes the process of converting maize into bioethanol more efficient, so there are higher ethanol yields and reduced water and electricity consumption. The cost savings have encouraged some of America's ethanol plants to adopt it commercially, but most people would agree that the reductions in environmental damage still don't make this form of bioethanol a sustainable option.

Biofuels made from food crops have fallen from grace both because of environmental concerns and because of fears they compete with food production. If our future requires both more food and more fuel, then using the same crops for both is clearly going to be a problem. A popular alternative to bioenergy from food crops is to use woody plants such as poplar trees, willow trees, switchgrass and miscanthus (elephant grass). There's a lot to be said for a source of biofuels which is abundant, needs fewer inputs on the farm and doesn't compete with food production. The US Department of Energy even hopes that these could replace the equivalent of 30% of America's petrol consumption. However, the process of turning woody plants into biofuels will have to be much more efficient before it becomes economically feasible. If you want to turn wood into liquid fuel and not simply burn for electricity, trees have come up with some challenges.

It is the cellulose in woody plants which is fermented into biofuels, and this is much harder to extract than the sugars and starches we use from food plants. The cell walls of woody plants are stubbornly resistant to being broken down, so biotechnologists have come up with a wide range of strategies to try and make fuel from cellulose an economically-viable option. One culprit for the difficulties in cellulose extraction is lignin, which gives structure to plant cell walls. There are various options for transgenic plants which are easier to convert into biofuels, such as reduced lignin, modified lignin, or increased cellulose. Rather than modifying the cell wall, we could get better at breaking it down. Part of the process of making biofuel is to break down the cell walls with enzymes from bacteria, and scientists have introduced genes for these enzymes into plants.

It looks promising that transgenic plants could produce higher yields of cellulose, and that converting the cellulose into fuel will take less energy than it does at the moment. Still, practical challenges aren't the only hurdles to overcome for woody biofuels. However efficiently you produce energy from wood, this technology will only bring environmental benefits if it's used wisely. The dangers are visible when you look at our current use of wood to create electricity. Each year, Drax power station in the UK imports millions of tonnes of wood pellets from forests in the USA to burn for energy. Emissions from shipping, along with the impact we're having on forests on the other side of the world, have to be factored into this equation. It may not be on cropland, but you have to grow the trees or grasses somewhere.

Certainly not everyone is convinced by using grasses and trees for energy. A recent World Resources Institute report concluded: "Cellulosic biofuels (sometimes referred to as 'second generation') may use crop residues or other wastes, but most plans for these biofuels rely on planting and harvesting fast-growing trees or grasses. At least some direct competition with food is still likely because such trees and grasses grow best and are most easily harvested on relatively flat, fertile lands—the type of land already dedicated to crops."

In fact, Senior Fellow at the World Resources Institute Timothy Searchinger summarised the report as: "Do not grow food or grass crops for ethanol or diesel or cut down trees for electricity."

GM trees which produce high yields of wood are already out there. On 9th April 2015, the Brazilian National Technical Commission on Biosafety (CTNBio) approved the commercial use of genetically modified eucalyptus trees. The company behind the technology, FuturaGene, stated in their press release that the approval represents 'the beginning of a new era for sustainable forest management by enabling the production of more fibre, using less resources'. The trees can be used for paper and bioenergy, and are ready to harvest in five and a half years rather than seven. This fast growth is thanks to a single gene, introduced from the common laboratory plant Arabidopsis. The regulators were convinced by the data from years of field trials addressing concerns such as gene flow, leaf-litter decomposition, and the composition of honey made by local bees.

The approval did, of course, attract outspoken opponents, and a consortium of organisations from around the world has led a highly-organised campaign against the trees. Shortly before the approvals were announced, a thousand women from the Brazil Landless Workers' Movement occupied FuturaGene's operations and ripped up seedlings. One of their arguments is that local bee keepers risk losing the international markets for their honey. This seems unlikely, and for environmentalists interested in food miles I'm not sure that having international markets in Europe is necessarily something to aim for. It is, however, an interesting reminder of the diversity of stakeholders in the debate.

Opponents have also questioned whether, given the complexity of trees and their interactions, we know enough to design adequate safety tests. For example, trees generally spread pollen further than crops do, potentially increasing the chance of gene flow to native relatives. This isn't an issue in Brazil as eucalyptus has no wild relatives there, but it could be a sticking point if the technology is to be approved in different parts of the world.

Brazil's eucalyptus plantations cover millions of acres, and the trees have already been bred for fast growth. It seems ironic that both sides believe that they are protecting biodiversity – FuturaGene by needing less land to produce the same amount of wood, and its opponents by protecting the environment from 'dangerous GM trees'. Could they both have got it wrong? Perhaps the correct question to ask is 'can we think of a better way to produce energy than vast rows of non-native trees?'

One obvious alternative is to use a plant which is already being grown for another reason. An appealing option is agricultural waste, and genetically-modified yeast could be used to turn by-products such as straw, corncobs and sawdust into biofuels. Also on the horizon are GM plants which clean up environmental pollutants, as we'll explore below, and there's excitement about generating energy with plants that have cleansed brownfield sites.

It's a fascinating issue, and once again GM is just a small part of a much larger discussion. There is definitely potential for GM techniques to tackle some bioenergy challenges. However, the best way to use plants to meet our energy needs, and indeed whether we should use them at all, is still up for debate.

Some critics of land plants as a source of energy are looking towards algae, which has the major advantage of only taking up small amounts of land for the reactors it is grown in. Professor Rod Scott from the University of Bath is working to develop strains of microalgae suitable for biofuel production. He described the benefits: "To provide 50% of the USA's fuel requirements with corn oil, you'd need 846% of the available crop area in the US, which is clearly impossible. That falls to just 2.5% with microalgae."

Far from needing prime agricultural land, algae will grow using waste products which are currently discharged into the environment. They can use municipal wastewater and carbon dioxide released from power stations, and there is the potential to grow them in contained environments. It might also be possible to make different biofuels with different techniques. Oils, carbohydrates and proteins produced by algae can be converted into biodiesel or bioethanol, or anaerobic digestion of algae by bacteria can produce methane. You can use large species of algae, such as kelp and other seaweeds, or single-celled microalgae.

The prospect of using microalgae to produce biofuel has been around since the 1970s when an energy crisis led the US Department of Energy to investigate the possibilities. It's certainly possible: at the 2012 Texas Mile land speed event, a motorbike powered by algae-derived diesel reached 96.2mph. We're back at a common problem though, that it simply isn't economically viable to produce algal biofuels on a large scale. As Rod said: "At the moment we use 90 million barrels of oil – over 14 billion litres – every single day. You can fiddle around with a flask of new fuel and think we're doing quite a good job, but we use a staggering amount of oil."

Microalgae does have the advantage that it can be grown in contained bioreactors, which drastically reduces a primary concern of many opponents to GMOs – the risk of escape into the environment. The question is whether these bioreactors will ever be able to produce the quantity of biofuel we hope for. Open pond systems are perhaps more feasible, but they take up more land, and will we be able to keep the risk of escape acceptably low?

Faster growth rates and tolerance to contaminants are the kind of characteristics which might be needed to make algal biofuels viable. However, these are also the characteristics which could give algae a competitive advantage if it escaped into the environment. Again, the issues aren't specific to algae which have been genetically modified, but they are of fundamental importance when assessing the safety of algal bioenergy.

Another challenge is that many algae species naturally produce toxins. Algal biofuels will therefore have some safety standards to pass, ensuring that nothing used in the process is harmful to human health.

For an idea which was first explored in the 1970s, algal bioenergy has a disappointingly long way to go. Hope is not gone though, and new technologies could bring success. Recent advances in genomics mean we can identify which genes produce different characteristics and design entirely new plants or organisms specifically for biofuels.

Just as we have a long history of growing our fuels, we also have a long history of growing our medicines. Like the energy industry, the pharmaceutical industry has moved to other sources and now might start moving back. Therapeutic proteins, edible vaccines and antibodies for immunotherapy could all be produced using genetically modified plants.

The first pharmaceutical protein made in plants was human growth hormone in transgenic tobacco. This experiment took place in 1986, and exciting steps forward over the following years included an experimental vaccine against the hepatitis B virus. Nobody is particularly keen to smoke vaccines, so progress now is being made with transgenic food plants. Clinical trials have taken place for edible vaccines produced in lettuce and potato – given my childhood fear of needles, this is something I can fully support.

For those less squeamish, there's also interest in making injectable vaccines by extracting compounds from GM plants. Crop plants could be used to produce vaccines, although there is also interest in ways to avoid the potential concerns of growing vaccines in an open field. One system which could be contained in a controlled environment is hairy root cultures. These roots can grow by themselves in bioreactors without the rest of the plant, as 'factories' to produce high-value molecules such as vaccines, industrial enzymes, cosmetics and food additives.

The origin of hairy roots is nature's genetic modification. In the wild, DNA can be transferred from a soil bacterium to a wounded plant. This bacteria, *Agrobacterium rhizogenes*, is related to the *Agrobacterium* species used in lots of genetic transformations. It can cause roots to grow abnormally and produce molecules used by the bacteria as food. Professor Milen Georgiev, a biotechnology researcher from the Bulgarian Academy of Sciences, explained: "In fact, all we are doing is just copying a natural phenomenon. This is an amazing example from Mother Nature of how a small organism is able to 'colonise' more sophisticated organisms, such as plants."

Hairy roots are easy to culture because they grow indefinitely, just like a tumour. They also have the advantage of genetic stability – theoretically every root line is generated from a single cell. As transgenic organisms, they avoid many of the concerns with GM. Transgenic hairy roots are cultured in sterile conditions, preventing transgenic material from being released into the environment. There is also a high degree of purification, so no transgenic DNA will be present in the final product.

The major challenge at the moment is, unsurprisingly, how to produce hairy root cultures on a commercial scale. There are already many forms of bioreactor used for plants and microbes – often cylinders of glass or stainless steel, some of which are large outdoor vats while others are packed densely into a lab. However, these classical bioreactor systems aren't suitable for commercial hairy root production. New bioreactor designs are being developed though, and Milen strongly believes that in the near future we will have commercial use of these bioproduction systems.

"In 20-30 years we will need more food and fibres," he said. "And eventually more healing molecules of plant origin. From this perspective, plant in vitro culture, and hairy root systems in particular, offer an alternative. I really believe that my efforts, and those of all people in the field, will result in better management of hairy root-based processes in order to develop so-called 'green chemical factories' in the very near future."

Hairy root systems could also have the potential to clear contamination from soil, and some naturally degrade contaminants such as TNT. The use of plants to remove pollutants from soil or water is known as phytoremediation, and it could be a relatively cheap solution to a widespread problem. We produce a whole host of chemicals which are damaging to human health or the environment – industrial effluent, agricultural chemicals, textile dyes, pharmaceuticals – and have inherited some of the worst contamination from previous decades.

Polychlorinated biphenyls (PCBs), for example, were widely used in electrical equipment. Environmental concerns led many countries to ban them in the 1970s but their stability means they are still present. In the 1960s, Rachel Carson's Silent Spring described the environmental damage caused by DDT, and this was one of the factors which contributed to its ban for agricultural use. The trouble is that pollutants such as DDT and PCBs are very difficult to remove from soils because they're not water soluble. As a result, there are lots of expensive and environmentally-destructive ways to clean up contaminated sites. Soil is often moved to landfill or burnt in an incinerator, or it can be treated using techniques such as irradiation or microbial degradation. Plants have the potential to be cheaper and less environmentally damaging than conventional ways of removing pollutants from soil or water.

There are already examples of phytoremediation success stories using non-GM plants. The Gulf War left Kuwait with severe pollution problems, especially from oil spills, and since 1995 ornamental shrubs and trees have had a positive effect on oil-contaminated soil. Now biotechnology allows us to look further afield for genes which degrade chemical compounds. These genes can be isolated from plants, bacteria or fungi, and then introduced into plants which can be easily grown on the contaminated land. Theoretically, there's also the potential to combine phytoremediation with biofuel production.

Heavy metals are major environmental pollutants released by human activities ranging from mining to military operations. Over 400 types of plant have already been tested for their ability to take them up. Of course the result is simply that the heavy metal is contained within the plant, and the next step can be to recover the metals through burning the leaves.

Pollutants of a very different nature can be found in our everyday lives. How long have you ever been without using something made of plastic? Just from this morning I have an alarmingly long list: my alarm clock, my bedside lamp, the packaging for my milk and cereal, my shampoo bottle, buttons on my clothes, my toothbrush, my water bottle, my lunchbox, my recycling bin, parts of my computer and phone, my pen...

Given our complete reliance on plastic, there are multiple reasons to rethink the way we produce it. Currently, much of our plastic is made from oil or natural gas. Around 4% of world oil production is used as a feedstock to make plastics, and creating plastics is an energy-expensive business. The fate of waste plastic on land and at sea is another huge environmental problem. The millions of tonnes of plastic dumped in our oceans every year are likely to take hundreds of years to degrade. Light breaks plastic down into small fragments (otherwise known as nurdles!) which get eaten by marine animals, and albatrosses have even been known to kill their chicks by mistakenly feeding them bits of plastic. Birds become tangled in fishing nets, plastic bags, and the plastic rings which hold beer cans together.

We've come a long way in developing biodegradable alternatives to plastic, often from plants and other natural sources. For example, many bacteria naturally produce PHAs (polyhydroxy-alkalates) which have become popular for making biodegradable plastic. They are often blended with starch or cellulose from plants and have been used for plastics ranging from water bottles to coated paper. However, a major drawback is the expense. Synthesising PHAs using bacteria costs about five times as much producing plastics from petroleum. A promising alternative is to introduce bacterial genes into plants so they produce PHAs.

Switchgrass, the predominant tall grass on the North American prairie, is an obvious choice as it could be grown on a large scale on land which isn't suitable for food crops. Theoretically, transgenic switchgrass could provide a double crop of both bioethanol and biodegradable plastic. And what about a triple whammy – engineering a crop plant so that the leaves would produce PHA and also be suitable for biofuel production? The 'plastic' genes wouldn't be switched on in the edible bit of the plant but, technical challenges aside, this perhaps has a squeamish factor which means consumers may not be keen on this any time soon.

It is no mean feat to create an environmentally-friendly way to turn transgenic plants into plastic. Work is underway to make this viable on a commercial scale, but there are still hurdles to overcome. For example, to ensure maximum benefits of biodegradable plastics we need to think carefully about how to separate them from conventional plastics so they are disposed of properly.

As well as providing an alternative way to create materials that we already use, genetic engineering could also allow us to grow substances such as spider silk which we can't otherwise manufacture. In a game of materials Top Trumps, spider silk would be a difficult card to beat. This incredible material is stronger than Kevlar (which itself can be 20 times stronger than steel), 100 times stronger than ligaments and more elastic than nylon. The problem is how to manufacture enough of it to be useful for anything.

Whilst I hope for the sake of horror movie directors that someone manages to make a spider farm, there are lots of reasons why it's very impractical. It is very different to producing silk from silk worms, which weave large silk cocoons around themselves. Silk worms are the caterpillars of moths which have been bred for human use to the point where they can't fly. They can be kept at very high densities, but if you tried this with spiders they would simply eat each other. Farming traditional livestock is far more convenient, so scientists turned to safer ground.

Silk-producing goats were originally engineered by a Canadian company whose 'BioSteel' plans attracted media attention in the early 2000s. However, the company ran out of money before they could bring the product to market. The idea is now in the hands of academics, and a herd of 'spider goats' now lives on a farm run by Utah State University.

The goats look exactly the same as normal goats, as does their milk. The difference is a single silk protein, produced only in their milk. Currently the goats' milk is filtered and dried to extract the protein, which is then made into a fibre and pulled through a syringe to make a thread of silk. The trouble is that currently the goats don't produce enough of the protein for this to work on a large scale.

It's spider silk, not goat farming, which is the Utah State University group's passion, so they are also exploring the possibilities of producing silk with genetically modified bacteria, alfalfa and silk worms. Whichever method they use, it is always a challenge to recover the protein for use.

Dr Justin Jones, a researcher at Utah State University, explained the problem: "Until now, this process was thought to require use of expensive, noxious solvents, many of which degraded the strength and elasticity of the spider silk protein. This made manufacture of the silk too costly, its consistency unreliable and posed a danger to workers."

Instead, Justin and his colleagues used water as a solvent and applied heat and pressure to extract the protein. A key piece of kit in their successful experiment was a microwave. "We blew holes in about eight ovens before we perfected our method," he said. "But a microwave is a relatively inexpensive tool to produce a commercially scalable and environmentally friendly way of producing consistently high quality spider silk."

Spider silk could be a biological alternative to Kevlar, which is used in tyres, racing sails, body armour and modern drum-heads, and its elasticity makes it suitable for replacing ligaments or tendons. The team has even been given US$1.9 million funding from the US Department of Energy to investigate how to make lighter, more efficient vehicles using spider silk.

This fairly wacky use of genetic engineering reveals quite how important it is to consider each genetically modified organism on a case-by-case basis. Spider-silk goats developed by university researchers don't come with problems of 'superweeds', herbicide use and the corporate dominance associated with glyphosate-tolerant plants. In the last six chapters we've considered an incredible array of possible GMOs, and each one needs to be considered on its own merits and risks.

Chapter 11

Coexistence

In the early 2000s, my school friend used an interesting analogy to convince me of the benefits of GM crops. She pointed out that, if the Wright brothers hadn't been brave enough to take the first flight, we wouldn't have the benefits of air travel today. It was a good reminder that fear of the unknown can prevent progress, but the analogy falls down. The Wrights risked their own lives, whereas the release of GMOs is theoretically a risk borne by everyone. At the time, environmental groups were warning that there would be no going back – once we had released GM crops into the environment, we would face widespread 'gene pollution'. Plants which bred with GM crops would invade natural ecosystems.

Today, concerns about contamination of our food and the environment still remain. The three major fears are that GMOs will turn up in non-GM food; that GM crops will grow in natural environments; and that GM plants will breed with their wild relatives. Although the dire consequences I feared at the turn of the millennium haven't materialised, all of these situations have occurred.

GMOs turning up in non-GM food hasn't caused health problems, but it has led to serious economic losses. Contamination of food with unexpected ingredients is extremely widespread, hence the prevalence of allergy warnings such as 'made in a factory which handles peanuts'. Similar contamination by GMOs is therefore perhaps no surprise, and both transport and processing are often responsible. The problem can also start back on the farm, where GM seeds are accidently grown with non-GM seeds, or where cross-pollination occurs between GM and non-GM crops.

The issue is particularly pertinent for co-existence of GM crops with organic farming, as a painful 2014 court case demonstrated. Australian farmer Steve Marsh lost organic certification on over half his land when GM canola was alleged to have blown into his fields from his neighbour's harvest. Marsh then sued his neighbour, Michael Baxter, for US$85,000 damages. After a 3 year battle, the Western Australia Supreme Court ruled in Baxter's favour, though not before the pressure had destroyed his marriage. Baxter had been using standard harvesting practices, and the court deemed that he hadn't acted negligently. Instead, they ordered Marsh to pay over US$800,000 in court costs.

The devastating impact on the lives of these two farmers shows that this issue shouldn't be a fight between neighbours, but is a wider challenge for society to tackle. Individual organic farmers differ in their views about whether GM crops have the potential to help them achieve their organic goals or are fundamentally in conflict with their ideology, yet official guidelines are clear: GM crops can't be certified as organic. This means organic farmers are left vulnerable to contamination in parts of the world where GM crops are widely grown.

Currently, GM-free positioning is a powerful marketing tool for organic produce, and organic farmers are understandably reluctant to give that up. However, adopting a zero-tolerance approach has implications for farmers looking to grow GM. It's not obvious where we should draw the line in this conflict. Is it right that a farmer is allowed to grow a GM crop which threatens their neighbour's organic business model? Or is it inappropriate that organic farmers serving a niche market can dictate what other farmers can and can't grow? The Baxter v Marsh case prompted some scientists to criticise the organic guidelines as arbitrary, and question whether we can expect farmers planting GM crops to change their management practices to suit organic farmers.

One widely-accepted compromise is to allow low levels of GM material to be present in 'GM-free' foods. Governments around the world have set their own limits, as we will discuss in Chapter 16. Studies in Australia have indicated that a tolerance level of 1% would be sustainable in the long term, although some countries have set the limit as higher. There have also been recommendations that compensation schemes are set up for farmers who suffer economic losses from contamination in situations where nobody has been negligent.

Even if we decide that some contamination is inevitable and acceptable, the more concerning events involve crops which haven't been through the regulatory process. Although this is yet to cause a health problem, these varieties haven't been con-firmed as safe for human consumption and so these cases highlight possible risks.

The first case of an unauthorised GM line contaminating in-ternational food supplies occurred between 2001 and 2004. Syngenta inadvertently produced and distributed several hundred tonnes of an unapproved GM maize. The crop was planted over 150 km^2, which is about 0.01% of maize planted in the US, and was distributed as food. The Bt10 insect-resistant maize differed only very slightly from an approved variety and was later assessed as safe by US government scientists. Thankful-ly, the problem didn't escalate once it was detected, and no incidents of contamination with this variety have been recorded since 2005. Syngenta paid fines of US$375,000, imposed by the US Department of Agriculture, and was instructed to sponsor a compliance training conference.

This followed a contamination case in which a GM variety approved only for use in animal feed was discovered in the human food chain. StarLink maize revealed the economic risks of contamination, not to mention the negative publicity which continues today. Over 300 products contaminated with StarLink maize were recalled, which is estimated to have cost the food industry around US$1 billion. There was also an impact on exports. Japan is a major importer of maize from the USA, and its zero-tolerance approach to StarLink in shipments of grain reduced the value of US maize in 2000/2001. It's therefore not surprising that Starlink's producer, Aventis CropScience (which has since been acquired by Bayer), faced expensive lawsuits. These included a demand for compensation from farmers who didn't grow StarLink but still had the value of their maize decreased.

The US authorities which approved StarLink for animal feed have been criticised for not predicting this eventuality. Unlike crops such as wheat, maize breeds through cross-pollination, increasing the likelihood that StarLink pollen would find its way into fields of maize approved for human consumption.

Other contamination cases include GM crops being grown accidentally in a particular country where they weren't approved for cultivation. In 1999, some European farmers unknowingly planted oilseed rape with a low proportion of herbicide-resistant seeds. The seeds had come from Canada, where the GM variety was being grown commercially.

The contamination only came to light after the crop had been harvested and consumed, causing a great controversy. Friends of the Earth called for a criminal investigation into how the problem occurred, and for all UK field trials of GM crops to be stopped while the separation process was reviewed. In reality, there was no persistence of the GM variety once the problem had been identified, and no ill effects were reported. This is perhaps not surprising as the seed had been through regulatory approval elsewhere, and any environmental concerns surrounding herbicide-tolerant crops only apply if you actually use the herbicide.

What should we learn from these cases of contamination? That no consumer or farmer can ensure their land and food will be free from GMOs? Or that, as none of these cases have led to known damage to human health or the environment, the dangerous effects of contamination have previously been exaggerated? To me, the take-home message is an element of both. It's yet another situation where we need to take both societal preferences and science into account when planning a route forwards.

The next concern is that GM crops will 'escape' and grow in natural environments. Modern varieties of crops have been developed to thrive under optimum agricultural conditions, with a plentiful supply of nutrients and free from competition with weeds. As a result, they aren't good at surviving in natural habitats; you don't see wheat plants growing in forests. What you do sometimes see, however, are crop plants growing in semi-natural habitats. These can be on the edge of farmland or along transport routes. Lorries and trains transport grains over thousands of kilometres, and inevitably containers sometimes leak and seeds spill out. If the seeds find themselves in the right environment, they're often able to grow.

This is also an issue with non-GM crop varieties, but there can be an added concern if the plants are herbicide resistant. Railways and roadside verges are often sprayed with glyphosate, the most common herbicide to which GM crops are resistant, and so the feral plants have a competitive advantage. Herbicide-resistant GM plants have been found growing along railway lines in countries such as France and Switzerland which don't grow GM crops – these are seeds which escaped from a freight train not from a local field. Better handling could reduce this problem, for example having covers on train carriages containing grain, although we are unlikely to eliminate it.

The presence of feral GM canola growing along roadsides in North Dakota prompted scientists at the University of Arkansas to question whether the USA has adequate oversight and monitoring to measure the environmental impact of GM crops. Professor Cynthia Sagers from the University of Arkansas and her graduate student Meredith Schafer travelled 3,000 miles looking for canola at the roadside. They sampled every 5 miles along the highway, and found wild canola at almost half of their samples sites. This wasn't all GM, but they found that 83% was genetically engineered to be resistant to herbicides. Their tests weren't able to detect herbicide-resistance developed through conventional breeding, and the problem could be even greater if non-GM herbicide-tolerant canola is present too. It is herbicide resistance, rather than genetic engineering, which is the issue here.

One of their biggest concerns was the discovery of plants resistant to two herbicides. No canola with resistance to multiple herbicides was commercially available, leading Cynthia and Meredith to conclude that two commercial canola varieties must have interbred in the wild. This potentially makes weedy populations of canola harder to control.

Current farm practices could be contributing to the problem. Each year US farmers leave tens of thousands of acres of canola un-harvested, creating a large reservoir of seeds which can escape into the wild. Management practices will be key if we are to reduce the unwanted mixing of GM and non-GM crops, including isolation distances between the crops, and the way we harvest and transport grains.

This interbreeding of two GM varieties of canola is accompanied by findings that GM crops can breed with wild populations of the same species, or even with related weed species. Some of the key findings come from the work of Professor Allison Snow, a plant ecologist at Ohio State University. Throughout her career, she has maintained a concern that there may be ecological questions about GM crops which aren't being addressed, and has set about tackling them.

Her interest in GM crops began in the 1980s, when plants were being genetically engineered for the first time. "I was fascinated by the science, by the elegance of it and where it was going," she said. "But as an ecologist I also was a bit appalled at the lack of knowledge of ecology from the people that were developing the crops."

In 2002, Allison presented data from experiments on Bt sunflowers, and her findings helped shape my precautionary views at the time. Cultivated and wild sunflowers often breed together in the USA – they are, after all, the same species. In some conditions, this gives wild sunflowers a competitive advantage, and there was a fear that this would be the case if they bred with GM varieties.

The GM sunflowers in the experiment were resistant to caterpillars, and they could pass this resistance to their wild siblings by interbreeding with them. Experiments showed that reduced pest damage allowed wild sunflowers to produce more seeds once they had bred with Bt sunflowers. Theoretically, this extra flower power could cause a problem in the wild. The Bt sunflowers were experimental varieties, but Allison and her colleagues predicted that they would breed with wild and weedy populations if they were released commercially.

We don't, however, know what the impact would be if this happened and the gene spread in the wild. Allison said: "While it's obvious that a single gene can have a huge impact on plant reproduction in a natural setting, there are still a lot of unknown effects, like whether or not the weed could spread at a faster rate."

In the case of the sunflowers, it looks like we will never know whether any damage would have been done. The original study had received funding from the companies which developed the seed, Pioneer and Dow, and they didn't supply money or seeds for future projects. This followed their decision not to seek regulatory approval for commercial release of the sunflowers, and still today there are no GM sunflowers on the market.

Not only can crops breed with other plants of the same species, many crops are also able to breed with wild relatives. These wild relatives may be found in natural habitats or be farmland weeds.

Problems could occur either if the plant gains an advantage from acquiring the new gene or if the gene proves to be a disadvantage. If the new gene is an advantage then the native species could cause disruption in natural habitats by out-competing other plants, and could become hard to control. If breeding with a crop is a disadvantage, then a wild species of conservation concern could have its survival reduced.

Despite extensive research into gene flow from GM crops to wild relatives, there are very few cases where this is known to have occurred. This may be partly because GM is mainly limited to a few major crops which are often grown far from their wild relatives, at least at the moment. The most extensive example of hybridisation comes not from a major crop but from an experimental variety of creeping bentgrass.

Creeping bentgrass is commonly used as turf on golf courses, and a herbicide-tolerant variety was planted in Oregon as part of an experiment in 2002 and 2003. Despite efforts to prevent gene flow, GM pollen was carried over 20 kilometres on the wind. It pollinated non-GM bentgrass, along with plants of a closely related grass species, black bent. To make matters worse, a wind storm in 2003 also dispersed creeping bentgrass seeds.

The spread continued after the experiments had stopped, and in 2007 the company behind the GM bentgrass was ordered to pay a US$500,000 fine for failing to comply with US regulations during the testing. Failures included inadequate cleaning of equipment and a lack of the required buffer zones around the crop. Nearly 15 years later, herbicide-tolerant creeping bentgrass can still be found growing wild in Oregon.

This persistence is partly down to some features of creeping bentgrass biology. For example, germination tests showed that in some situations GM creeping bentgrass seeds can still grow after being buried in the soil for over 4 years. Like much of the work on GM bentgrass, this study was performed by Oregon State University weed scientist Professor Carol Mallory-Smith and her colleagues. Their studies in the years following the GM bentgrass release led them to conclude that, despite control attempts, we aren't likely to ever eradicate populations carrying the gene.

Carol doesn't believe that our findings about gene flow should necessarily prevent the introduction of GM crops, but this knowledge should be included in risk assessments. Information about gene flow is essential for regulatory approval, and also for designing management practices for GM crops. These need to be considered on a case-by-case basis, depending on the crop and the genes involved. Creeping bentgrass, for example, presents a high risk: it is wind-pollinated, has small seeds, and breeds with wild relatives. In contrast, some of the major crops don't breed with wild relatives, although canola is known to inter-breed with 40 weed species globally. The GM characteristic is also important: some may be a disadvantage to the plant in the wild, while others will bring benefits.

The results from studies with bentgrass are relevant when considering the introduction of GM wheat. Just as creeping bentgrass can breed with wild relatives, wheat has also been known to breed with jointed goatgrass, an agricultural weed. Like so many of the concerns with GM crops, in reality this issue is much wider. Weeds can also acquire herbicide-resistance genes from crops developed through conventional breeding, and we've already seen the movement of a non-GM herbicide-resistance gene from wheat to jointed goatgrass.

In fact, the best example of a problem caused by gene flow actually comes from a non-GM crop. Rice farmers around the world report problems with weedy rice, also known as red rice, which is a wild relative of cultivated rice. It can dramatically reduce yield when it grows as a weed in rice fields, yet its similarity to domestic rice means it's very difficult to find a herbicide to kill one but not the other. Herbicide-tolerant rice seemed to be a solution, and this has been developed using conventional breeding. Just three years after the herbicide-tolerant rice was released, it had bred with weedy rice to cause a resistance problem. This meant that a new technology which increased yields and profitability quickly became less effective, and it has even been withdrawn from the Costa Rican market following severe weedy rice infestations.

All these case studies reveal potential risks, even if the main problems we've seen so far have been economic. As a greater variety of GM crops are made commercially available, we will be faced with new decisions and new challenges. In some situations, the risks may be very low – for example, crops such as cassava, banana and potato are sterile. But, as the creeping bentgrass story demonstrates, other GM crops could be much more risky.

There are many management practices which can reduce the risk of genes escaping into wild populations, such as cleaning farm machinery, rotating crops to prevent problems with seeds left from the year before, and staggering planting times so that GM and non-GM crops don't release pollen simultaneously. Having a buffer zone around the crop can also be very effective, and for some crops that isolation distance only needs to be very small.

There have also been technological solutions proposed. Terminator technology, which could render plants infertile, caused an outcry because of the potential economic and social impact. Yet it could be a way of preventing GM plants from breeding with non-GM crops or wild relatives. Another option is that, theoretically, a gene could be introduced which causes a disadvantage to wild plants but isn't a problem in an agricultural setting.

We already have experience of managing co-existence with conventional crops, such as segregating edible oilseed rape varieties from those with industrial uses. Certainly contamination by GM crops is something we can reduce. However, we aren't ever likely to eliminate it. What we need to decide is how much it matters. In which situations can we ensure contamination and interbreeding remain at levels which cause no damage to health and the environment?

Whilst feral crops and wild plants containing 'escaped' genes don't necessarily lead to agricultural or ecological problems, one of the challenges is that we can't be certain when problems will occur. Our current knowledge is extremely valuable when assessing the risks and how to reduce them, yet we are still left with unknowns.

As Allison said, "It's not always the end of the world if a weed starts to become a lot more common after acquiring a new trait – there may be effective ways to manage that weed. You just can't make sweeping generalisations about genetic engineering, and knowledge from ecological studies like ours can help inform risk assessment and biosafety oversight."

In conclusion, the triffids are yet to invade, but we still need to proceed with caution.

Chapter 12

Human Health

On 18th September 2012, the UK's Science Media Centre received an unexpected tip-off about a brewing media storm. A French scientist had approached journalists with his latest paper on the health impact of GM crops, and asked them to sign a confidentiality agreement which prevented them from sharing the paper until the embargo was lifted. Embargos are used to give journalists time to write a story without the fear that someone else will beat them to it, and part of this can be asking independent experts to comment. A desire to avoid expert comment is enough to ring alarm bells, and so several journalists put in a call to the Science Media Centre.

Established with the aim of renewing public trust in science, the Science Media Centre helps journalists access scientific experts, particularly for controversial stories which can be surrounded by misinformation. This was exactly such a story, so they set about finding people to comment. Even without access to the paper, they could line up experts ready for when the embargo was lifted.

One of the people to help them in their mission was Dr Mimi Tanimoto, who at the time was running the UK Plant Sciences Federation. Without even seeing the paper, Mimi realised this could be a threat to the plant science community: "We knew it could be very damaging if inaccurate conclusions were released unchecked in the media, so I contacted people we could trust to judge how reliable the claims were. Even though most of us in the scientific community aren't used to the tight timescales of the media, we realised this was a situation where we needed to work fast. It wasn't hard to find people willing to drop everything and work with us on this."

Predictably, when the embargo was lifted there was cause for concern. Professor Gilles-Éric Séralini and his team at the University of Caen in France had concluded that rats suffered health problems both when consuming the herbicide glyphosate (which we discussed in Chapter 6) and when fed glyphosate-resistant maize. They reported that rats fed the GM maize were more likely to develop lethal tumours and experience severe liver and kidney damage than those fed on standard wheat, and also that rats developed tumours when their drinking water was spiked with glyphosate.

When Mimi spoke to her experts, their conclusions were remarkably consistent: "I sent the paper immediately to toxicologists and plant scientists, and they all agreed that the authors had drawn the wrong conclusions. This gave us confidence that we could spread a strong message to journalists and directly to anyone discussing the issue on social media."

Statisticians criticised the statistical tests, and expressed concern that there were simply too few rats to draw conclusions. Biologists pointed out that the type of rats used were very prone to cancer so weren't suitable for longer-term studies such as this one. Through Mimi, the paper also arrived in my inbox, and it was immediately clear that the journal's reviewing process had failed simply based on the poor quality of the language.

Criticism for the paper grew, and members of the scientific community called for Séralini to release all his data, not just the summaries he'd provided in the paper. Instead of releasing the data, however, he responded with claims that he was being dishonestly attacked by a lobby.

Whilst Séralini questioned the motivations of his opponents, doubts were raised about his neutrality. Funders of the Séralini lab have included Greenpeace, organisations promoting organic produce, and producers of detoxification products. A funding source doesn't necessarily dictate the outcome of a research project, but this very unusual group of funders suggests a biased starting point. Séralini was also promoting a book alongside the study, with the catchy title of *Tous cobayes!* – We're all guinea pigs.

A biased starting place was certainly Mimi's conclusion: "Séralini appeared to set out to prove that GM foods and glyphosate are damaging. So this was the conclusion he drew even though it didn't fit with the data."

The European Food Safety Authority (EFSA) reviewed the paper, and concluded that the study wasn't of high-enough quality to draw meaningful conclusions. The journal which had published the study received calls for the paper to be retracted, which it eventually was. It was later republished in a small environmental journal without going through the normal review process, a move which attracted widespread scientific criticism.

Years later, the conclusions of the paper still circulate as news on Twitter. It is these images which represent perhaps the most disturbing aspect of this story. They show a sad-looking Professor Séralini holding up a rat with a tumour larger than its head. In the UK (and indeed in many countries) it is illegal to keep laboratory animals in this kind of study alive with tumours above a certain size, much smaller than the ones depicted. Could it be that the rats were only kept alive for tumours to grow this big in order make a more striking publicity shot? Even if the conclusions in rats held true, nobody is suggesting that people eating GM maize will grow tumours as big as their heads. That clearly hasn't happened. But a picture, as they say, is worth 1,000 words, and the thoughts the picture conjures up don't necessarily need to be accurate.

Whilst Séralini has continued to publish papers claiming negative health impacts of GM crops, hundreds of other papers have concluded that there are no health concerns. Alongside this, regulatory bodies have concluded that all approved GM crops are safe, and there have been no confirmed cases of harm to human health. In Chapter 15, we will discuss the safety tests required during the regulatory process, which are both based on laboratory experiments and animal feeding studies. Animal feeding studies have been performed on a wide range of species, and no differing effect of a GM diet has been found in quail, rabbits or channel catfish.

Animal feeding studies are undoubtedly helpful when analysing the safety of individual compounds. It's relatively simple to feed compounds such as pesticides, food additives and pharmaceuticals to test animals, sometimes at higher concentrations than humans would be exposed to. These studies are also a valuable way to evaluate any proteins added into a GM plant – you can use bacteria to produce Bt toxins, and then feed these toxins to animals to determine whether they have a negative impact. However, the WHO is sceptical about how useful animal feeding studies are in the safety assessment of GM foods themselves. For a start, feeding only one food type won't provide a balanced diet, which is likely to be bad for the animal's health in itself. It's also hard (and arguably inappropriate) to get animals to eat excess quantities of the food, which would be a way of increasing any negative effects and so making them easier to identify.

Whether or not it is appropriate to use animals in studies which may have limited value is a moral issue, it hasn't changed the overall conclusion about the safety of GM food. In fact, there is widespread scientific agreement that we are yet to see any negative health impacts of eating GM foods. I can't possibly list all the governments and scientific organisations around the world which have made official statements about the safety of GM foods. These range from the Union of German Academies of Sciences to the Vatican's Pontifical Academy of Sciences, and are perhaps topped by the World Health Organisation. It states: "No effects of human health have been shown as a result of consumption of GM foods by the general population in countries where they have been approved."

A common theme for statements of safety is a lack of difference between GM and conventional foods. The US Food and Drug Administration, for example, states that: "FDA has no basis for concluding that bioengineered foods differ from other foods in any meaningful or uniform way, or that, as a class, foods developed by the new techniques present any different or greater safety concern than foods developed by traditional plant breeding."

None of this evidence is proof of the safety of GM food: science can demonstrate the presence of danger, but can't confirm its absence. Professor Nigel Halford from Rothamsted Research explained: "The difficulties of applying traditional toxicological testing and risk assessment to whole foods, GM or otherwise, makes it pretty well impossible to establish absolute safety. The fact is that very few foods consumed today have been subject to any toxicological studies, yet they are generally accepted as being safe to eat."

The studies we've done so far are also only evidence that the current GM foods are safe. Whilst it seems that GM foods aren't inherently dangerous, it is still possible that individual foods could come with a health risk. The WHO states that: "Individual GM foods and their safety should be assessed on a case-by-case basis and it is not possible to make general statements on the safety of all GM foods."

In reality, more GM crops are used for animal feed than for human consumption. As no health risks have been identified from humans eating GM foods, it's no surprise that the same is true of eating meat, eggs or dairy products from animals fed GM feed. Regulators assess the quality and safety of meat, milk and eggs from livestock fed on new crop varieties, and they have never identified any problems. The scientific evidence about safety of animal products has also been independently reviewed by non-governmental agencies such as CAST (the Council for Agricultural Science and Technology) and EFSA (European Food Safety Authority). They have come to the same conclusion about safety, and studies haven't even detected transgenic DNA in meat or milk.

Despite this evidence, fears are widely aired about health risks from GM foods. By far the most wide-spread claim is allergenicity, and some people have even blamed GM foods for the increase in allergies seen in our society. However, this claim doesn't stand up, not least because allergies have also increased in parts of the world where GM crops aren't widely available. The WHO's conclusion is: "No allergic effects have been found relative to GM foods currently on the market."

Again, this verdict covers only GM foods which are currently available, and it is impossible to make sweeping statements about all future varieties. Could GM foods trigger life-threatening allergic reactions? In theory, absolutely. If you transferred a gene for a protein known to cause an allergic reaction, then your new plant would put allergy sufferers at risk. And what a risk it would be – it is hard enough to be allergic to peanuts when they are easy to recognise, in their pure form at least. If your peanut allergy could be triggered by different fruits, for example, that would leave you with serious problems. Of course in reality nobody would put known allergens into different species, but the problem could hypothetically still occur if we don't know which proteins trigger reactions. For this reason, tests for allergenicity are part of the approvals process for new GM foods.

The case of a GM soybean with increased allergenicity has already been widely reported, with the problem caused by introducing a gene from Brazil nut trees. The soybean was meant for animal feed, but the allergen problem was identified and the crop never made it to market. A similar situation halted a decade-long project to create a pea which is resistant to attack by pea weevils. Australian scientists transferred a gene from the common bean into the pea, and found that it created an allergic response in mice. Even though the protein doesn't cause allergic reactions when it comes from the bean, its structure is subtly different in the pea and so it was an allergen in the new crop. To make sure such a product never reaches consumers, there is a strict procedure of laboratory tests, skin prick tests and food tests with volunteers to identify problems.

There's also another side to the story of GM foods and allergies, in that GM foods could potentially be developed with reduced allergenicity. Early-stage studies so far have included removing some of the allergenic proteins in tomatoes, peanuts and cow's milk. Nothing has made it to the regulatory stage, but in 2015 scientists published results of the first successful clinical trial of a GM food with reduced allergenicity. Twenty one volunteers with apple allergies ate some genetically modified apple, and approximately half of them suffered no symptoms whatsoever.

The potential health benefits of GM crops go beyond allergenicity, and some have already been studied. A particular focus has been mycotoxins, which are naturally-occurring chemicals produced by certain moulds. They are widely present in grains and other foods, and at high doses are carcinogenic and extremely toxic for humans and animals. Unsurprisingly, this means that countries often have a legal limit on mycotoxin contamination, which has a particular impact on trade. Developing countries may export grain with lower mycotoxin levels, and retain those above the limits for themselves.

There's evidence that Bt crops can reduce the problem. When caterpillars damage a plant they can make it susceptible to attack by moulds. By preventing pest damage, some Bt crops reduce mycotoxin levels compared to conventional crops. Bt crops can't eliminate mycotoxins, and factors such as soil management and timing of the harvest also influence levels. It does, however, show that GM foods have the potential to affect our health in unexpected ways.

As well as safety, it's also valuable to compare the nutritional values of GM and conventional foods. Environmental variation means this isn't an easy study to do, but evidence so far suggests they are broadly equivalent. Similar weight gain has been seen in animals fed GM and non-GM maize, though several studies have shown better growth rate and productivity of animals fed Bt grain. This is likely to be because of lower mycotoxin contamination on the insect-resistant grain.

Whilst it appears that there are no inherent dangers of GM foods, an area perhaps worthy of more consideration is the health effects of pesticide and herbicide residues on crops. As we saw with glyphosate in Chapter 6, this is often an extremely controversial topic. Chemical residues are regularly found in our foods, though this in itself isn't necessarily a problem if the concentrations are low. In fact, a toxicologist's mantra is 'the dose makes the poison'. Paracetamol is a prime example of a substance which can cause liver failure and even death if taken at a high dose, while is an effective painkiller if taken moderately. Even water and oxygen can be toxic if too much is drunk or absorbed.

What this means for agricultural chemicals, such as pesticides and herbicides, is that even those which are poisonous if they're consumed at high quantities are entirely safe if consumed in the trace amounts generally found in food. We now have very sensitive techniques for detecting pesticide residues, and just because we can detect pesticides in food it doesn't mean they are harmful.

Part of the pesticide approval process is to ensure that any chemical residues remaining in the crop won't be harmful to humans. A safe upper limit is set, and food and drink are monitored through a surveillance programme to make sure this isn't exceeded. However, it won't surprise you to hear that this process isn't particularly effective in many parts of the world, and independent studies have regularly identified food which contains chemicals in concentrations exceeding the official maximum limit.

While huge amounts of work have been done on the safety of GM foods themselves, data is decidedly lacking about chemical residues. There's currently no evidence that glyphosate-tolerant maize or soybeans have more herbicide residue than conventional crops. Levels of residue vary greatly and this type of analysis needs to be done on a case-by-case basis. We also need to consider whether other agrochemicals are found at different levels on GM and conventional crops. The potential risks of pesticide residues will no doubt be an ongoing topic of discussion, and GM crops need to be part of that.

Overall, the molecular biology techniques used for food safety analysis have seen staggering advances. The challenge won't be to detect differences in new varieties of crops; it will be in deciding which of these differences are important. Changes are likely to be found in new varieties, whether or not they're developed using genetic engineering, but this doesn't necessarily mean they are a cause for concern.

As we design and analyse such safety tests, it's worth remembering that eating is a risky business. A tragic reminder of this came in 2011 when organic beansprouts contaminated with *E. coli* killed 31 people in Germany and infected thousands more. Health authorities issued warnings against the consumption of cucumbers, lettuce and tomatoes before finally tracing the outbreak to beansprouts from a single organic farm.

For some people, even uncontaminated foods are a lethal risk. If peanuts were made through genetic modification, no regulators in their right minds would approve them.

It's not just nature which has provided dangers, so has conventional breeding. Examples include the Lenape potato, which was withdrawn when it was found to have dangerously high levels of solanine. Naturally found in potatoes, tomatoes and other members of the nightshade family, solanine is poisonous in high doses.

In fact, living itself is a risky business. We know stress is bad for you, so if this is making you uneasy I advise you not to search 'health risk of bread' or 'health risk of hand sanitiser'. And definitely not the health risk of alcohol, cars or computers.

There are certainly risks associated with exaggerating dangers, as the nocebo effect can make them a self-fulfilling prophecy. The opposite of the better-known placebo effect, the nocebo effect occurs when people suffer harm from a harmless substance. It's commonly known in medicine – if you read about the side effects of your medication then you are much more likely to suffer from them. News reports can do this too. In New Zealand the reports of negative effects of a thyroid treatment rose 2,000-fold after a new formulation of tablets were released. Nothing other than the size and colour of the drug had changed; yet news stations around the country had reported side effects of the new formulation and this had become a self-fulfilling prophecy. This response occurs to food as well as medicines, and studies have concluded that some patients' food intolerance is simply a nocebo.

Let's not forget that a UN report about the nuclear disaster in Chernobyl concluded that: "Persistent myths and misperceptions about the threat of radiation have resulted in 'paralyzing fatalism' among residents of affected areas". In fact, the report labels mental health as "the largest public health problem created by the accident". This damaging psychological impact was partly caused by a lack of accurate information – just one of the many ways the devastating impact of this tragedy could have been reduced.

Chapter 13

Patenting Life

Photos of 1930s farming show us a different world. They depict men moving hay with pitchforks, and ploughs being pulled by heavy horses. Yet there is a similarity to today's situation which took me by surprise: unlike their parents, many 1930s farmers didn't save seed from their maize harvest to replant the following year. Instead they bought new seed each season from commercial traders. In the early 20th century, research into new maize varieties meant that the seed farmers could produce themselves didn't deliver the same yields as seeds they could buy.

No patents were needed to make it profitable to sell seeds, because there was a feature of plant breeding that kept farmers coming back for more. The highest maize yields came by crossing two inbred lines to create hybrid seeds. If you breed from these hybrid plants the second generation doesn't have the same characteristics, so farmers don't save these seeds from their harvest. This trend for hybrid seeds has a similar effect on farmers to 'terminator technology', without the general outcry.

Hybrid seeds share some similarities with labradoodles. Cross a Labrador with a poodle and you get a calm and sociable companion which doesn't shed much fur. Most breeders stick with this reliable recipe: take two inbred lines and create a single generation which is both desirable and predictable. Inbred plant lines have the major advantage that they don't come with the medical conditions which often plague pedigree pets.

This trend was accompanied by innovations in plant breeding. In the 1950s scientists started using chemical and radiation mutagenesis to increase the genetic diversity in their breeding stock, as we saw in Chapter 4. Initially, the traders were family businesses selling varieties which had been developed using tax payers' money at universities or research stations. However, selling seeds proved profitable, and soon there were larger businesses on the scene, some of them doing their own research. By 1965, over 95% of the American land used for maize was planted with hybrid seed, and now hybrid seeds dominate agriculture in many parts of the world. Decades before the advent of biotechnology it was already worthwhile to buy new seeds every year.

Although the maize market was dominated by hybrid seeds, farmers still saved their seeds for many other crops. Breeders couldn't charge royalties on these seeds, until new rulings came into force. In Europe, Plant Breeders' Rights were born in 1961 with the formation of the International Union for the Protection of New Varieties of Plants (UPOV), and they came into practical use in the 1970s. Today they give exclusive rights to use a variety for 25 years. By granting breeders of new plant varieties an intellectual property right, UPOV aims to encourage 'the development of new varieties of plants, for the benefit of society'. It now has member countries from six continents.

Plant Breeders' Rights are the primary way that European breeders receive royalties on their new varieties, and in the UK these royalties are collected by the British Society of Plant Breeders (BSPB). About a third of the money is reinvested into breeding programmes, and BSPB's chief executive, Dr Penny Maplestone, sees this as absolutely critical: "The industry would just not function if we didn't have effective intellectual property.

You need a system to get a return on your investment – without protection there can't be any breeding."

Developing a new crop variety is an extremely expensive process – in the UK it can cost in excess of £1.5 million a year to run a competitive wheat breeding programme. The cost has often relegated smaller breeders to niche crops, such as peas and beans. These crops may be important nutritionally, and as nitrogen fixers, yet we are in a vicious circle where a lack of suitable varieties makes farmers less likely to grow the crops. Penny explained: "The size of the market means royalty income is hardly anything. It can sustain about one and a half breeding programmes, and they are permanently on a knife edge. It's a concern that these smaller crops are strategically important, but nobody can afford to do the breeding."

Penny's biggest concern in the UK is that new varieties aren't being developed fast enough. At the current rate of innovation, she doesn't see us as being able to tackle future challenges of increasing productivity in a changing climate. We need to go faster, and Penny considers an effective intellectual property system as a critical factor in enabling us to increase the speed of innovation.

Penny started her career in academic research, and believes that public-private partnerships can increase the speed of innovation, bringing in knowledge from academic research. Companies don't have enough money from royalties to be able to do speculative research, which is more likely to be done in academic institutions. This more fundamental research will be essential for major advances in breeding.

In fact, Penny is worried there's a misconception that every-one who works in the industry is 'out to control the food supply'. At the BSPB, she has spent almost two decades working with plant breeders, and she finds the complete opposite is true: "All plant breeding companies, whatever size and shape, want to make a profit because that's what companies need to do to exist. But I think breeders all genuinely believe that they can make a difference to food and farming, and that's why they are in the business."

Plant Breeders' Rights aren't completely equivalent to a patent, as they come with exemptions. For example, varieties protected by Plant Breeders' Rights can be freely used in breeding programmes (something which has a lot of support from the breeding community). Although Penny is confident that Plant Breeders' Rights will continue as the mainstream system in the EU, there has been an increasing interest in patents. She explained: "Patents are a more expensive business to get into, so people will maybe look into this when they've got something special they made a very significant investment in. They may feel justified in seeking stronger protection in order to benefit from the monopoly you can get through a patent."

Patents may be a focus for criticism of biotechnology, but patents for living organisms were already being granted over a century before the first transgenic plant was created. Louis Pasteur, the French scientist whose inventions include treating milk and wine through pasteurisation, received a patent on isolated yeast for brewing beer in 1873. Intellectual property on plants likewise has a long history. At a banquet at the Royal Horticultural Society in 1899, British judge Lord Justice Lindley is reported to have said: "Without being a prophet, I seem to see before me a vista of patent hybrids! What a treat for the patent lawyers! And what an accession of work for Her Majesty's Judges!"

His prediction certainly appears to be coming true – lawyers and judges have their work cut out with determining exactly what can be patented. It was originally more clear-cut: for example, it has been possible to patent asexually-reproducing plants in the USA since 1930. In parts of the world, you might notice some roses are sold as patent protected, with 'Asexual reproduction of this plant is prohibited' on the label. For the duration of the patent, which is 20 years in the USA, you need a licence to take cuttings or reproduce the plant asexually in any way. There's nothing to stop you from using seed or pollen to reproduce the roses sexually, and in some countries there are exemptions for personal use.

Agricultural patents now go far beyond this. Debates rage about exactly what can be patented, and there is widespread disagreement about whether these patents are supporting or hindering innovation. To add to the confusion, patent laws vary hugely around the world, and the intellectual property systems can be very different. Canada for example doesn't allow patents to cover plants or animals.

Whereas Plant Breeders' Rights allow the protection of plant varieties, many patents cover a particular trait (or characteristic) which can be introduced into many different varieties. Much of what plant breeders do isn't patentable because it doesn't meet the criteria of being novel, non-obvious, inventive and useful. However, even conventional breeding can lead to patents, and recently there have been two high-profile plant patents awarded in the EU. One of these was for a tomato with a reduced water content, which makes for easier processing, and one for broccoli which is high in anti-carcinogenic glucosinolates. These cases have been hotly contested, and their award follows a decision by the highest court of the European Patent Office that plants or seeds obtained through conventional breeding methods are patentable.

So far, patents have been much less common than protection through Plant Breeders' Rights. To date approximately 120 patents involving conventional breeding methods have been awarded in Europe, whereas in 2014 Plant Breeders' Rights protection was offered to 3,633 varieties from the UK alone. Many more patents have been applied for, however, and time will tell whether this is a growing trend.

Increasingly, patents and Plant Breeder's Rights are becoming interlinked. Plant Breeder's Rights cover plant varieties. But Bt cotton, for example, isn't a plant variety – there are many different cotton varieties with the same Bt genes added. For the developers to cash in on their invention, it is the Bt genes which are patented.

Even though few companies have the resources to develop and patent their own GM technology, they can still add existing GM technology into their own seeds. For example, Monsanto has licenced the Roundup Ready technology to over 200 seed producers. The practice of licensing means that a single modification ends up in crop varieties suited to a whole range of conditions.

When farmers buy patented or protected seeds they sign stewardship agreements, in which they commit not to replant seeds from their harvest or to sell these seeds for planting by anyone else. Such agreements aren't unique to GM. BASF's herbicide-tolerant Clearfield crops, developed through conventional breeding, come with an agreement which states "…saving seed to plant next year's crop is not allowed". BASF has successfully taken a father and son from Louisiana to court for replanting Clearfield rice over multiple years.

As with other industries, patent infringements and subsequent court cases are rife in agriculture. In 2015 Bayer Crop-

Science successfully sued a Chinese company for selling counterfeit pesticides, and Syngenta has done the same over its insecticides.

The seed giants which have gone to court to defend their seed patent rights include DuPont, Syngenta, BASF and Monsanto. Each year, over 325,000 farmers buy patented seeds from Monsanto and sign an agreement not to replant them. Between 1997 and 2015 Monsanto filed 147 lawsuits against American farmers, though only nine cases have gone to a full trial. The millions of dollars collected in these cases have been donated to youth leadership initiatives such as scholarship programmes.

In a legal battle lasting six years, Canadian farmer Percy Schmeiser came up against Monsanto. The case attracted the world's attention as 'David vs Goliath' when Monsanto sued Schmeiser for growing GM Roundup Ready canola without a license. Percy claimed the seeds must have ended up in his field by accident, yet this seemed somewhat unlikely given that crop which was approximately 95% Roundup Ready was grown on over 1,000 acres. His other line of arguing was that he hadn't actually exploited the gene by using glyphosate, and so Monsanto didn't have proprietary rights over the plants. The Supreme Court disagreed, and ruled against Schmeiser in 2004.

Numerous American farmers have come up against Monsanto in high profile seed-saving cases. Vernon Hugh Bowman lost his case in 2013; in 2010 a jury ruled against Mitchell and Eddie Scruggs, who were deemed to owe US$6.3 million damages to Monsanto; and Homan McFarling lost multiple appeals in relation to the seeds he'd replanted in 1999 and 2000. To trump it all, Kem Ralph from Tennessee was sentenced to 8 months in prison for selling saved seeds, lying under oath and burning seeds which were evidence.

One thing all these cases have in common is that the seeds were knowingly and intentionally planted. Farmers have never been sued for accidentally growing patented crops, which could legitimately happen through contamination. Still, fear of this possibility drove a coalition of organic farmers, seed producers and anti-GM groups to take Monsanto to court. The long and complex case of Organic Seed Growers and Trade Association et al. versus Monsanto ended in what the growers considered to be a partial victory. It legally bound Monsanto to its existing assurances of not taking action against growers whose crops might inadvertently contain traces of patented genes.

There are situations where saving proprietary seeds is legal, as long as the grower pays a royalty fee. In the UK, the BSPB is responsible for collecting royalties for farm-saved seed, and this amounts to approximately £10 million per year. Despite the royalty, it is still cheaper for farmers to save their seed than purchase new seed. This, conversely, means less income for the breeder, so there is a business case for creating hybrid seeds which can't be saved and replanted. BSPB's Penny Maplestone explained: "There is incentive for plant breeders to produce varieties that won't be farm saved, and that's one of the reasons that a lot of them are looking at hybrids. There are technical advantages of hybrids as well, but the farm-saved seed element is obviously a business factor."

The issue of patenting goes well beyond the seed industry – our food system contains a long chain of patents. Just as your mobile phone was made with a multitude of components sourced from different countries covered by a multitude of patents, your can of soup will be the result of a wide range of patents. The creator of the first known commercial tomato soup, James Huckins from Boston, Massachusetts, began canning soup in the late 1850s and received a patent for his recipe. Even growing the ingredients may involve patented techniques – in 1986 a patent was granted to Campbell Soup Co for its method of culturing tomatoes. So soup can be created with ingredients grown from protected seeds, using patented tools and patented agricultural inputs. And that's before patented machines with patented components were used to put it into its patented packaging.

Plant patents have many and varied opponents, from the National Farmers' Union to Greenpeace. Two key concerns are lack of transparency and high cost. Patenting has created a new minefield for breeders, who are at risk of infringing patents without realising. It can take 18 months for the US patent office to publish successful patents, for example, and it's hard to know if you're using patented techniques in this time. The high costs of filing patents, and perhaps defending them in court, is yet another expense which can most easily be borne by large companies. Multi-nationals also have the habit of buying up smaller companies and acquiring their patents. There's also a concern that not all countries have a breeders' exemption, so patented plants can't always be used in new breeding pro-grammes.

For these reasons, it is possible to use patents to stifle inno-vation, yet they also have the potential to promote it. Professor Richard Jefferson has risen to the challenge of using patents to enable innovation, and he told me about his journey towards founding The Lens, a website hosting most of the world's patents and turning this information into knowledge.

His interest began with patented lab techniques, which are widely used both by conventional breeders and biotechnologists. Such patents can be a source of income for companies and universities alike, and they mean that the effect of patents on agricultural innovation starts well before seeds are commercialised.

For developers of GM crops, patents on laboratory techniques can take complex navigation. There are over 1000 patents just surrounding the use of *Agrobacterium* to introduce genetic modifications into different crops (something we covered in Chapter 4). This complexity is yet another factor which plays into the hands of the larger companies: they are the ones with the resources to understand the situation. A desire to provide everyone with access with the information they need for innovation was one of the things which spurred Richard into action.

Now a social entrepreneur, Richard began his career in the lab. His interest was in creating new techniques, and he was one of the scientists behind the first field experiment with a genetically modified crop. On the 1st June 1987 he planted GM potatoes just outside Cambridge in the UK, as part of a study on the biology of genes in the field. Only later did he discover that his group had beaten Monsanto to the world's first GM field trial by just one day, something which remains a little-known fact.

His first major contribution to molecular biology came right at the start of his career with the GUS reporter system, a technique he developed as a grad student in Colorado. GUS reveals where in an organism a particular gene is activated. This was revolutionary: for the first time it was possible to monitor exactly what happens when foreign genes were introduced into an organism's DNA.

Alongside his successes in the lab, however, he became increasingly disillusioned by the hierarchical nature of science, which was at odds with the egalitarian society he'd experienced growing up in Santa Cruz. This all came to a head when rumours got out about the potential of GUS. A senior professor contacted one of Richard's colleagues and asked for some samples. Richard saw that GUS could be used to 'enrich the Old Boys Network', and knew he had to challenge the system: "I just blew up. I said 'nope, he's not going to get it any faster than anyone else'. I was not going to let that happen."

Instead, Richard persuaded a friend to sneak him an address list of around a thousand plant molecular biologists. He wrote a huge user manual and filled padded envelopes with tubes of GUS reagents, then posted them to researchers around the world. He sent it to people in the public and private sectors, ensuring that it reached scientists in Bangladesh at the same time as those in Berkley.

Within just weeks of receiving GUS, scientists were able to genetically modify soybean for the first time, and it is now one of the most widely-used tools in plant molecular biology. Richard ultimately patented GUS, and created licensing terms which promoted technology sharing. This success got him thinking: "What if we had an institute which actually designed tools to solve problems in the field, and designed them to fit the hands of the people who really needed to solve the problem? If you make an effective and affordable method available to everybody, everybody will use it. That was the genesis of Cambia."

Cambia is a non-profit institute set up to enable innovation and promote change. As founder and CEO, Richard's aim is to facilitate sharing of information, knowledge and wisdom, with a particular focus on marginalised communities. It started with a project to facilitate the development and sharing of rice biotech methods. After four years of Richard 'bugging' the Rockefeller Foundation, he secured funding to go to every rice biotechnology lab in Asia. He described the early days: "At the time Cambia didn't exist except as letterhead, so Rockefeller made the cheque out to me personally. They gave me $100,000 and said 'now make this happen'."

And that's exactly what he did. He moved from the Netherlands to Australia to avoid being crippled by jetlag, and began his regular trips to Asia. It was a successful time: he met his wife in the Philippines, and developed and shared new transgenic rice technologies around Asia and beyond.

Meanwhile, the agricultural industry continued to restructure into a few large players who could use their patents to deter others from accessing valuable research tools. The result as Richard saw it was public concerns over new technologies, and missed opportunities to use these technologies for public good. He began to realise that we needed more than science in the quest for sustainable agriculture. He explained: "Science is often just a tiny piece of the jigsaw puzzle. We realised how critically important intellectual property was going to be, not just for saying 'no' but as a way to signal where people have skills and capabilities."

Richard and his team built the world's first tool for searching the full texts of patents and made it freely available. Patent Lens, which has now been superseded by The Lens, became very widely used. Richard next set his sights on understanding 'innovation landscapes'. Who is doing the work? What are they doing? What's the science? What are the patents? He first analysed the landscape of the techniques surrounding *Agrobacterium*, and published this as an open access document.

To put this together, Richard and his colleagues had to read and understand the patents, something he described as "a bit like having root canal surgery every week for fun". But it gave everyone a tool to navigate the patent landscape. This kind of knowledge can be extremely difficult to access unless you have a sizeable pot of money for paying lawyers to find out whose patents you might be infringing. Once again, innovation becomes extremely difficult for all but the largest players.

The Lens is designed to allow anyone to access the kind of information which businesses pay millions of dollars to understand. By providing knowledge which can guide decisions, Richard works to support entrepreneurs without deep pockets. He said: "We need to reduce the cost of doing business intelligence to zero – we need to make it easy for people to find good partnerships and discover blockages in intellectual property. By using The Lens, anyone, anywhere can make a clear path to product development."

Ultimately, Richard believes that we should be proud of the core technology of genetic engineering, but not proud of its implementation. Instead of tackling this problem by vilifying the few businesses which dominate, he has dedicated the last three decades to supporting more players from both the public and private sectors in creating solutions to the world's challenges.

The largest companies in the seed business have a complex web of cross-licensing for GM technologies, although licensing deals haven't always been reached without a fight. In 2013, for example, Monsanto asked US courts to dismiss its claim that DuPont infringed its patents, in exchange for DuPont dismissing its claim that Monsanto uses monopoly power to stifle innovation. Instead, DuPont now pays Monsanto a royalty for technology used in herbicide-tolerant crops. Perhaps making friends with your competitors is a good way to reduce their concern over your dominance. Licensing agreements will become even more complex with the rise of GM crops containing multiple new genes.

A particularly bitter debate has unfolded around CRISPR, the genome editing technique we considered in Chapter 4. Researchers from two separate public institutes in the USA believe they were the inventors, and they entered into a battle of contested patents. It's rare that public institutions are willing to rack up the lawyers' fees involved with taking a patent case to a hearing, but with licencing fees set to yield many millions of dollars, the stakes were high. In February 2017, however, the Broad Institute in Cambridge, Massachusetts, received the verdict it had been waiting for. Three patent-court judges ruled that its patent application was distinct from the University of California's, meaning that the Broad's patents still stand.

The CRISPR patent battle is far from over, and hundreds of other patents have now been filed around the technology. The explosion in the use of CRISPR and the time taken to process the patent applications means there could be trouble ahead for companies using the technology. Therapeutics companies using CRISPR have been fuelled by venture capital funding exceeding a hundred million dollars, and licensing fees could cost them dearly. It is unlikely that academic researchers would be sued for patent infringement, partly because the outcome of lawsuits is often a share in the profits, which amounts to nothing if an academic is doing basic research. Still, such patents risk being a barrier to research, and some people see the CRISPR patent debate as evidence that universities shouldn't file for patents.

Perhaps even more than the other issues we have considered, patenting is a complex challenge with no obvious answer to the question of what impact does it have on agriculture overall. Patents can be used to make investment in innovation worthwhile, or they can be used to limit access to valuable technology. Sometimes patent rights have led to expensive battles, whereas patent owners in other situations have given technology for free to benefit the poorest farmers (as we saw in Chapter 7 with the story of golden rice). The issue is once again by no means limited to GM crops, and it is intimately connected with concerns about corporate control.

Chapter 14

Corporate Control

Most of us care far more about our health and the environment than about the profits of companies we don't even work for. The trouble is that some employees of multi-national companies don't share our priorities – something that was clearly demonstrated by the 2015 revelation that VW had been cheating in emissions tests.

As with most multi-nationals, you don't have to look very far in the agricultural industry to see instances of bad behaviour by people at the top. For example, BASF was one of the eight companies fined by the European Commission for price fixing of vitamin products throughout the 1990s. In 2002, a senior manager at Monsanto directed an Indonesian consulting firm to bribe a high-level official in a bid to avoid environmental impact studies. This wasn't an isolated incident, and Monsanto has also admitted to other similar bribes. Dow and Monsanto are among the chemical companies to have had to pay compensation in relation to Agent Orange, having initially denied that dioxin contamination of their defoliants caused veterans' medical problems.

These make shocking (though perhaps not surprising) stories. From sole traders to FTSE100 executives, you will always find people willing to engage in dishonest practices. Laws, regulations and audits haven't eliminated these, although they do make the cases relatively rare. Arguably, the far larger problem is the legal, systemic consolidation of companies into a few giants, and the inequality it can produce.

Professor Richard Jefferson is very concerned that the seed market is dominated by a few big players, but also points out that society as a whole is responsible. We built the rules for companies, and following these rules allows companies to shape the food system to their benefit. Our society is set up for businesses to put profit over social benefit. He said: "Companies are profit maxmisers, obliged to consider the value to their shareholders. They have a vested interest in keeping high barriers to entry into markets. Does it make them evil? Not really – it makes them realists, because we have set up institutions which celebrate that." We built the rules for companies, they are following the rules.

Our current situation follows decades of transition.

The 1970s was a time of change for the seed market, as increased protection of intellectual property made seeds an attractive proposition to larger companies, including the agro-chemical industry. Over the following decades, patents related to genetic engineering also provided strong incentive for mergers and acquisitions, and many smaller biotech companies were acquired by multi-nationals. Today the Big Six companies dominate the global seed market: Monsanto, DuPont, Syngenta, Bayer, Dow, and BASF.

The road towards monopoly has been characterised by the key players buying up smaller companies and competitors. This hasn't finished, and in 2015 Monsanto made a bid to take over Syngenta. Syngenta made its position pretty clear in its letter of response to Monsanto's CEO Hugh Grant – "Syngenta neither sought nor welcomed an acquisition proposal" – but it still gave a glimpse of the world as Monsanto sees it.

It is perhaps ironic that Monsanto's next move wasn't to attempt another takeover, it was to accept a takeover bid. Bayer's US$66 billion planned acquisition of Monsanto has given rise to antitrust concerns amongst regulators – the companies currently have some competing products even though Bayer's main agricultural offering is chemicals and Monsanto's is seeds. The two companies are working hard to tempt Donald Trump with promises of investment and jobs following a merger, and we should learn later in 2017 whether competition laws will prevent the deal from going through.

This wasn't the only major merger bid in 2016. In August, US regulators signed off a deal between Syngenta and ChemChina. ChemChina has been buying up businesses around the world, and if other regulators allow it through, the US$43 billion takeover of Syngenta will be China's biggest ever foreign deal. An impending merger between Dow and DuPont is also facing intense regulatory scrutiny and opposition from farmers groups.

Monopolies in the seed industry can go hand-in-hand with agrochemical monopolies. Even though the Roundup herbicide (glyphosate) is off patent, Monsanto is by far the biggest seller. It continues to come out with new formulations, and it has tied the purchase of Roundup Ready seeds to its own Roundup herbicide. Syngenta has done the same thing by bundling a non-GM barley variety with a fungicide, demonstrating that the power of the seed industry is much more deeply entrenched than GM.

If you ask your friends 'who makes GM crops?' the chances are they will all say Monsanto. Monsanto is indeed a major player which holds very important intellectual property, though it's not clear why Monsanto and GMOs seem to have become so completely synonymous. Whilst its competitors don't share the limelight, they certainly share Monsanto's practices.

All these businesses are part of a commercial seed market which is set up to meet the needs of large-scale farmers in the developed world. Seeds which are optimised for the developing world aren't such a good commercial prospect, and so poorer farmers generally don't see the benefits. Even the developed world may not see their interests being met in the long term. Crop varieties are being developed for today's world, and may not be suitable for future challenges. The seed market lacks diversity, and this doesn't set us up well to deal with climate change. We also can't rely on being able to continue with current farming practices. Rising oil prices are just one of the reasons that we may not be able to sustain the level of inputs which current seeds are optimised for.

It isn't only farmers buying GM seed who are reliant on the large seed companies. Even in Europe most farmers return to the same seed companies each year to buy the high-yielding hybrid seeds. It's pretty hard to fight back against this situation, and there certainly doesn't seem to be any political will to do so. However, interest is growing in alternatives. There have been attempts to introduce the favourable characteristics found in hybrid seeds, which can't be saved, into open pollinated varieties, which can be saved. There are also companies which specialise in 'heirloom' seeds, older varieties not developed by the big seed companies. The challenge in these situations is to create varieties which are the same quality as those produced by the multi-nationals. Understandably, farmers generally choose what will give them the biggest profit, and currently that tends to be the commercial varieties.

Interestingly, there has been a rise in open source seed development, just as there has been a rise on open source software. The model is basically the same: seeds are freely available, and can be shared, sold, and reproduced. It's fine to modify and improve the line of seeds, as long as it remains free for others to use. A fascinating twist in the tale of open source seeds is that some GM technology is coming off patent, and the gene for glyphosate-tolerance already has. There are complexities to the story, such as maintaining regulatory approvals, but there are also some advocates of creating open source biotech seeds. Whilst Monsanto is the antithesis of the open source movement, its technology could become part of it.

The ethics of the seed industry is a fascinating and emotive issue, yet it is only a very small aspect of corporate control. From farm machinery and inputs, through packing and distribution, right to the marketers and sellers, our food system is run by big business. Food goes on a complex journey from farm to plate, and there are lots of people who want shares of the profit along the way.

Downstream from the farm, we see the same picture of domination by a few large businesses. Nestlé, for example, employs around 335,000 people in 194 countries. In 2014 the Nestlé group had a total equity of over US$70 billion, just a few million more than the GDP of Luxemburg… for farmers, there is often little choice of who to sell to. In the Brazilian soybean market, for example, 200,000 farmers attempt to sell to five main commodity traders. Globally, there are around 25 million coffee producers, yet almost half of coffee roasting is carried out by four firms.

As far as retailers are concerned, supermarkets have a very large percentage of the food market. In Australia, for example, just two supermarket chains have over 70% of the market. In most countries, you can count the number of major supermarket brands on one hand, yet they have thousands of suppliers and millions of customers. The scale of the issue becomes even clearer when you look at the number of brands which are owned by the same parent companies. Restaurants such as McDonalds are also big players, and the size of some chains is even bigger than meets the eye. Pizza Hut and KFC, for example, are just two of the chains owned by Yum! Brands.

For the multinationals involved at all stages of the supply chain, a major concern is that size can mean power – the ability to shape the food system to their own needs. Companies can fix prices, influence government legislation, determine what food is grown and how it is packaged, and affect consumer choices. The power of a few businesses has left us without a diverse, competitive market.

At the same time, it's important to think about benefits this system is providing. In some ways, these economies of scale have served me as a consumer well. With a quick trip to one shop I can easily purchase a staggering selection of food using a relatively low proportion of my earnings. However, is the price I'm paying in any way a reflection of the environmental cost of my food? And how many of the people who brought this food to the shelves are paid a fair price? For the sake of my health, did I buy the right food? Often, our food system promotes some unhealthy choices. Is it everyone's right to make these choices, or a sign that our needs aren't being met correctly?

Inequality is also an inherent feature of a system with a few powerful players, with wealth flowing to those at the top. The power of supermarkets over suppliers, for example, often means that farmers are unable to demand a fair price. Abuses of this power have taken the form of charging a fee to get onto the supplier list, or returning unsold goods to the supplier even if they then can't be sold on (a contribution to food waste as well as a problem for farmers' income streams).

Not only do we have unequal benefits with the greatest benefits going to those at the top, we also have unequal distribution of wealth globally. The price wars in British supermarkets have led to tropical fruits being cheaply available to western consumers, while the farmers and workers are trapped in a cycle of poverty. The campaign group Make Fruit Fair calculated that farm workers received just 4% of the retail price of a pineapple while the retailers got over 40%. The British dairy industry has seen a steady stream of farmer bankruptcies because supermarkets have been paying so little for milk that it can be impossible to make a living. The issue made headline news in the UK in 2015 with farmer protests, which included bringing cows into supermarkets to meet the shoppers. The situation improved slightly as a result, though the problem is far from solved.

Obesity is one symptom of a system which treats food as a profitable commodity rather than a human need and right. Whilst I can't blame my excessive cake consumption entirely on corporations, they do influence the population's health. This can be through their product formulations or with tactics such as advertising campaigns designed to encourage poor health choices. In fact, the current state of affairs is win-win for corporations. The parent food companies sell some foods which are healthy and some which are the opposite – you can promote obesity whilst simultaneously cashing in on it. Unilever, for example, not only owns Ben & Jerry's but Slimfast too.

The history of the tobacco industry shows what a powerful motive profit can be, even when lives are at stake. Advertising to children and lobbying against policies which promote health are signs that Big Food risks taking a similar path to Big Tobacco.

These issues are a reminder of the irony of British supermarkets holding anti-GM positions (even if those tend to have been relaxed now). Because of negative consumer perception, 27 of the top 30 European retailers don't stock GM products, and until recently many didn't sell meat from animals fed GM feed. A large part of the anti-GM sentiment was related to corporate control, yet supermarket bans on GM were designed to keep customers buying from these huge corporations.

There are clearly many issues which need to be addressed, though is some of the obsessive anti-corporation attitude blinding us to the benefits which the big businesses can bring? After all, I know I happily criticise the pharmaceutical industry for failing to publish the results of unsuccessful drug trials, yet I have no qualms about benefitting from their medication. Likewise, it is tempting to condemn the profits of large businesses whilst enjoying the cheap food they can provide. When we think about how to reform the food system, we need to think about what we want to keep as well as what we want to lose. By focusing on our favourite targets such as Monsanto, do we allow many of the culprits to get off too lightly? And do we question the big players whilst failing to question the system which created them? Perhaps we need to rethink our entire attitude to profit and economic growth.

Studies focussing on the economic impacts of GM food have often assumed that social benefits will automatically follow. However, just because a technology brings economic growth, this doesn't necessarily mean it will increase wellbeing. In some cases, the economic benefits of a technology such as GM crops will no doubt bring benefits to the individual. For the poorest farmers, small improvements to their financial situation can make a big difference to their lives. Likewise, economic growth can have social benefits on a country scale. If you look at the poorest countries, measures of health and wellbeing rise rapidly with increasing income per person. However, this relationship breaks down for the richer countries. Health and social problems are strongly related to income inequality, but only weakly related to average income. It gets to a point where more money doesn't make things any better. The implications of this are again a far wider debate than GM, but it's a reminder to think carefully about what economic benefits actually mean.

Part of the anti-corporation feeling which surrounds the GM debate isn't related to inequality but to credibility. Many critics don't trust claims from agri-businesses that GM foods are safe. However, just as there are vocal doubters of corporations' claims that GM is safe, there is growing criticism of corporations which aggressively push the message that GM is unsafe. What is spun as corporate responsibility can be perceived as deceitful advertising.

The international restaurant chain Chipotle discovered this the hard way when it announced it was going GM free. The media's response included a Washington Post headline of 'Chipotle's GMO gimmick is hard to swallow' which accompanied an article criticising its attempts to make indulgence wholesome. The Wall Street Journal pointed out a contradiction in Chipotle's customers avoiding GM foods whilst merrily adding extra sour cream to their 1,000 calorie burrito bowl. Chipotle founder Steve Ells has been quoted saying that the decision is about the chain's mission of 'food with integrity'. How a food can otherwise lack integrity isn't clear, although the campaign-style information on Chipotle's 'G-M-over it' webpage confirms that GM doesn't meet its integrity criteria.

Big businesses certainly have the power to shift the industry away from GM if profits demand it, as was elegantly demonstrated when McDonalds and others decided they wouldn't use GM potatoes. The decision was triggered by fears of consumer rejection rather than genuine safety concerns, yet their announcements spelt the end for existing GM potato varieties. This could be seen as consumer's power over corporates, or corporate power over agriculture. Or maybe it's both. Either way, it's a defeat for science.

Just because corporations excerpt control over our food system, it doesn't mean they control the mouths of scientists in the way that many campaign groups would have us believe. Pretty much every scientist who has spoken positively about GM crops has been accused of being secretly paid by Monsanto (it's always Monsanto). Professor Alison Van Eenennaam is just one of the many scientists who have experienced very open abuse as a 'lover' of these agrochemical giants. Now an American citizen, she was born in Australia, and an internet meme circulated calling for her to Go Home. Supporters of the US anti-biotech movement spread the message that she 'needs to get out of my country!'.

The water was muddied in 2015, however, with the misadventures of the University of Florida's Professor Kevin Folta. A long-standing and vocal advocate for GM technology, Kevin advertised himself as an independent scientist. The problem came when a Freedom of Information request led to the publication of thousands of emails he'd exchanged with Monsanto in a relationship that began two years earlier. Kevin and his friends at Monsanto shared the view that misinformation is blocking progress in agriculture, that people are amazingly gullible when it comes to misinformation about GMOs, and that, in Kevin's words, "the world is officially nuts".

Inevitably, when the New York Times posted the emails online, Kevin was subjected to the modern phenomenon of trial by Twitter. There were scientists lamenting that he had damaged their reputation of independence and environmentalists expressing views that he was a 'raving lunatic for Monsanto'. A photoshopped image showed his head on a baby's body being fed from a bottle of 'Monsanto Money'. On other social media sites the more offensive comments included assertions that his dead mother would have been ashamed of him.

The situation got so bad that, a year later, he wrote: "[Anti-GMO communities] have ridden me to the point where I was ready to quit science, where I wanted to just disappear. I have been personally harassed to the point where I sat in a plane, off to another talk, and hoped it would crash in a ball of flames. No kidding."

It was more than just 'cozyness' with these giants that caused the storm, it was money. Monsanto is listed on University of Florida's website as having given over US$1 million, and Syngenta over US$10 million. Much of the criticism focused on a donation of US$25,000 that Monsanto had made to support the University's Talking Biotech program, which ran workshops to train scientists in how to engage the public on agricultural biotechnology. As criticism poured in, the University publically announced that they would redirect the funds to the on-campus food bank. Make no mistake; there is no question of Kevin benefitting personally from Monsanto's money. Still, this connection was seen as too much.

In an email to science journalist David Kroll, Kevin shared his side of the story: "I've learned one huge lesson about this as the naive scientist that says yes to any opportunity– It is not what it is. It is what it looks like. If I had this to do over again I'd absolutely call this a relationship, a corporate lover, whatever they want. Going forward, even a cursory interaction will be documented and displayed for the world to see. That's where I need to be right now. The problem I had was never disclosure or transparency. This is a ruse to tear down my credibility by creating a false narrative of hidden secret funds and a clandestine relationship with a company that ignites hate in activists sworn to oppose modern agriculture."

His initial naivety seems quite remarkable given he was no stranger to death threats because of his point of view. But then, academic relationships with industry are very common, and he felt he was in the clear because no corporations funded his scientific research. This, perhaps, caused the lack of foresight. In response to the question did he or anyone at his University ever think accepting industry funds might be a bad idea, he replied: "No way. My program existed, it was good, but funding limited. When they asked me if funding would allow me to do more, I absolutely said YES. Are you kidding me? If you teach science and someone offers a way to let you do more of it, you do it. The program was funded by honoraria that could have gone to me personally, but instead went into the outreach program. So decent bucks from a deep pockets company? Of course!"

Did he make a mistake, at least in the way he cultured his image? I think that has been proved by the way this has played out. It's not just trust for him that has been eroded in this escapade, it's also trust in university scientists. However, that in no way means his views were bought by Monsanto or that all the criticisms are fair. It is clear from reading the emails that his relationship with Monsanto developed from a shared belief that the GM debate had gone wrong. Their correspondence didn't show his views moulding to Monsanto's as he closed in on the money.

As is common in these cases, the real issue was side-lined in favour of mud-slinging. Lots of researchers receive funding from agribusinesses – this is not only legal, it is often actively encouraged. The economic downturn caused governments to tighten their belts and research funding was an obvious area for cuts, so inevitably researchers look for alternative funding sources. It's also entirely appropriate that businesses should want independent studies and should pay for them. For some, however, this relationship has become altogether too cosy. It is by no means the case that companies can simply pay academics to come out with the desired conclusion (although there are past instances of the tobacco industry doing just that). But, for example, how does the option of industry funding influence which directions academics choose to follow? And is there inadequate government funding for the kind of studies which industries might rather we didn't do? Perhaps we need to move to a new funding system whereby some types of study aren't funded directly by industry. Instead businesses could pay into communal pot which is allocated by an independent body.

There may not be enough public money spent on agricultural research, but in some cases it has led to agricultural innovations which don't rely on big businesses. As a result, fears of corporate control don't necessarily apply to all GM crops. Although Monsanto's glyphosate-tolerant crops show how GM can perfectly support an agricultural system dominated by big business, in Chapter 17 we will consider some crops developed for public good rather than profit.

From research funding to company mergers, I have just scraped the surface of some enormous issues. However, a familiar conclusion holds true: GM is just one part of a much wider issue. As a European consumer, I see evidence of corporate power every time I go to the shops. That corporate power exists regardless of whether or not the products are genetically modified.

Chapter 15

Regulations

"The New GMO Apple Is a Health Hazard, but the USDA Approved It Anyway"

Sound familiar? This headline greeted the approval of the non-browning Arctic apple in 2015. It echoes a common sentiment around GMO approvals: governments say yes to things we don't need, don't want and don't trust. The standard 'government ignores danger' news stories imply that the approval was given on a whim, ignoring any evidence. In reality, there are years of safety data behind every approval. These news stories have one thing right though – the evidence is generally served alongside a large dose of politics.

Lawyer Greg Jaffe has watched the GM regulation debate unfold over the last three decades, ever since he chose it as the topic for his junior year thesis. He now works at the Center for Science in the Public Interest in Washington DC, advising regulators around the world – from the USA to Burkina Faso. He's looking to help countries build strong, but not stifling, regulatory systems that independently review GM crops for safety.

The challenge is to set up regulatory systems which protect humans and the environment, whilst allowing safe applications to move forward. Greg said: "We can have safe applications of this technology and unsafe applications, no different from any other technology. The question is how to determine if an application is safe, and how to do that in a way that is fair."

Rebecca Nesbit

Around the world there are many different takes on what fair and effective regulations should look like, and some countries don't have regulatory systems at all. Most developed countries have comprehensive regulations, covering environmental and food safety issues. These systems do, however, vary in their effectiveness. "I think I could come up with problems in all systems," Greg said. "There's always room for improvement."

Top of the developed world's 'room for improvement' list is arguably Europe. The EU's approach, widely seen as the most stringent in the world, is based on regulating the process by which the GM plant is produced. Whether the crop is herbicide resistant or insect resistant is irrelevant, what matters is whether the techniques used to create it are classed as genetic modification. Any plants created with GM techniques are in for a bumpy ride – clearing the EU system often takes many years, especially if the crop is going to be grown and not just imported.

Although few EU countries grow GM crops, over 40 varieties are licensed for import for food and animal feed. A company wanting approval to import its GM product must first submit an application to the national authority, and the next step is a risk assessment by the European Food Safety Authority. Then, after a vote by EU member states, the European Commission will draft a decision to accept or reject the application. This will be assessed by the Standing Committee on the Food Chain and Animal Health. If they agree with the EC's decision then the process is complete, if not it goes to yet another level and is assessed by the council of ministers.

Like many, Greg is concerned that politicians vote without having to justify their decision. He explained: "The European system to me is totally indefensible. The decision on whether to approve a crop is based on a political vote which is not at all reviewable by a court of law. They don't have to base their vote on any particular evidence."

The evidence they can choose to accept or ignore includes a safety assessment which asks the question of whether a GM variety is as safe as its non-GM counterpart, both for the environment and human health. It looks for immediate effects, such as health concerns, and possible long-term effects such as insect resistance.

The tests for any effect on health involve both lab experiments and animal feeding studies. For example, rats are fed the GMO for 90 days to look for any adverse effects, and chickens are part of 42 day feeding trials to investigate whether there are nutritional differences. Meanwhile in the lab, the introduced protein is exposed to enzymes from intestinal fluids. There is also analysis to see whether the introduced compounds are similar to known toxins or allergens.

Like most regulatory authorities, the EU bases safety testing on the concept of substantial equivalence. The aim is to identify any differences between the GM food to its conventional counterpart. Tests involve a range of biochemical comparisons looking for any toxicological or nutritional differences.

In addition to safety tests to approve crops for import, the EU also deals with approvals for GM crops to be grown. This is a more extensive process, with additional tests to investigate environmental impact. Most crops end up being held in the regulatory pipeline for at least five years, and not all make it through. The regulatory process adds between €10 and €20 million to the cost of developing a new GM variety – an amount which only the largest agribusinesses could be expected to cough up.

Appropriately, any GMO that has been authorised for cultivation continues to be monitored. This can be specific data related to that particular GMO's risk assessment, and will include regular screening of published scientific literature. The approvals must be renewed at least every ten years.

Since the EU's regulatory system was adopted in 1990, approximately 20 applications for commercial cultivation have been submitted. Only one new GM crop has been approved for cultivation since 1998 – a potato with altered starch content. It was authorised in 2010, but lasted just two years. It was grown on a very limited scale before its developers, BASF, suspended cultivation because of the difficulties involved with operating in the EU.

This fated potato, Amflora, was never intended to be eaten – it was designed with favourable properties for industrial processes. It has to be said that potato isn't what first springs to mind when you think of the plants required to make paper, yet its starch is widely used in paper pulp and glue. Starch typically makes up about 8% of a sheet of paper, and Amflora was designed to make life easier in the processing plant.

The years Amflora spent in the hands of the regulators were despite the fact that it seems about as low risk as a GM crop can be. It wasn't going to be eaten, and potato doesn't come with the same risks of gene flow as other crops – it's vegetatively propagated and has no wild relatives in Europe.

Professor Nigel Halford from Rothamsted Research described the regulatory challenges: "Amflora was bogged down in this process for over ten years, which is farcical. One of the reasons this took so long is the EU's beginning point of substantial equivalence. With Amflora you don't have substantial equivalence – you have deliberately changed its composition."

Comparing GM varieties with existing varieties in the hunt for substantial equivalence is the basis of risk assessment strategies across the world. Exactly how we could move beyond this and reliably assess the risks of a new crop isn't clear. Suggestions include a demonstration that the new proteins in the GM plant have a history of safe use, even if they had previously come from a different source.

In 2005, the European Food Safety Authority (EFSA) had already deemed that Amflora was safe for human health and the environment. However, when the European Commission issued a draft decision to authorise the crop in 2007, the EU member states were divided and didn't sign it off. The Commission went back to request another scientific opinion, adding more years to the process. In 2009 EFSA confirmed its judgement that Amflora was safe. The European Commission then authorised the potato, to the dismay of campaign groups. In the other camp, there was cautious optimism amongst biotech supporters that the feeling in Brussels may be warming towards GM crops.

The story, however, didn't end there. BASF had already pulled production by the time Amflora was dealt its final blow: the withdrawal of its approval. This followed an earlier mistake by the European Commission (EC) which didn't allow countries a second chance to vote. The European Food Safety Authority (EFSA) had published an updated scientific assessment in 2009, at which point the EC should have submitted a new proposal for approval. However, the Commission went ahead and adopted the original version, in violation of its own procedures. Hungary challenged that decision in court, with support from Austria, France, Luxembourg and Poland. The courts ruled in Hungary's favour, and Amflora had its regulatory approval withdrawn not for safety reasons but because of a failure to follow the correct procedure.

Rebecca Nesbit

GM crops are far more likely to gain approval on the other side of the Atlantic, though the processes are still complex. In the USA, there are three main agencies involved in GMO regulation. When novel plants are going to be released into the environment, the Animal Plant Health Inspection Service (APHIS, which is part of the US Department of Agriculture) conducts risk assessments to look for potential impacts on agriculture and the environment. Its role is to protect agriculture from pests and disease, including making sure that all new genetically engineered plant varieties won't become 'pest plants'. The Environmental Protection Agency (EPA) regulates any GM crops and products which contain pesticides, looking at potential risks to health and to the environment. The US Food and Drug Administration (FDA) ensures GM food products fit with existing food safety guidelines, testing whether the GM food poses any greater risk than its non-GM counterpart. The approval process is similar to the procedure for new pharmaceuticals.

Part of the approvals process is a public consultation, where anyone can submit a response either for or against the GMO. In reality, a large portion of the opposing comments are often based on a template 'fill in the blanks' letter. By law, all comments have to be considered, but that doesn't mean this tactic is effective. An APHIS spokesperson explained: "It is the substance of comments rather than the quantity that impacts our decisions."

Food and food ingredients derived from genetically engineered plants must meet the same safety requirements that apply to food from conventionally bred plants. Again, the aim is to show substantial equivalence between the GM and non-GM varieties. However, the FDA's consultation process is voluntary, and not the mandatory pre-market approval process which you might expect. The company which developed the GM crop produces a safety assessment, to determine whether it could be

239

toxic or allergenic when eaten and to compare the levels of nutrients in the GM plant to traditionally bred plants. FDA scientists then evaluate this safety assessment and also review data from other sources, such as scientific literature.

The European system may have been criticised for being too stringent, but the FDA's system has attracted criticism for not being a reliable test of safety. The critics don't just include environmental lobby groups – the Center for Science in the Public Interest (CSPI) is amongst the NGOs concerned that FDA policy doesn't assure the safety of genetically engineered foods. Its 2003 report 'Holes in the Biotech Safety Net' argued that "the enormous potential benefits from GE crops and foods will be fully realised only if FDA's regulatory system is significantly upgraded and enhanced."

Over a decade later, the CSPI still has concerns about the system. Greg shared these in a testimony before the Senate Agricultural Committee, telling the hearing: "While the current genetically engineered crops grown in the United States are safe and beneficial, the federal regulatory oversight system needs significant improvements to ensure safety for future products and to provide consumers with confidence about their safety."

A major issue is that, although all GM crop have been through the consultation process, at the moment this is voluntary. Greg's recommendations include that before any GM crop is turned into food, the FDA should have to formally approve that the crop is safe for human and animal consumption. There needs to be a mandatory pre-market approval process that is open to public participation and review.

As for environmental risk, the CSPI has concluded that the EPA does a reasonably good job regulating pesticide-producing plants, although there are areas that need improvement. For example, the EPA could also improve its oversight of Bt crops after their release, and ensure that farmers comply with refuge requirements.

On the environmental front, the US Department of Agriculture's assessments have the strange loophole that it doesn't regulate all genetically engineered crops, only those classified as 'plant pests'. Bizarrely, this means that plants made using *Agrobacterium* are regulated, whereas those made using a gene gun are not, simply because *Agrobacterium* is classed as a plant pest. In his testimony to Senate, Greg pointed out this flaw: "Developers and USDA spend significant resources determining that a GE crop is not a plant pest when they could use those resources to analyse and address real impacts from GE crops, such as development of resistant weeds and pests, or gene flow to wild relatives and non-GE farms. It is difficult to find any credible scientists who think adding one or two new genes to a domesticated crop would turn it into a 'plant pest'."

While the USA lets the marketplace deal with any possible social or economic concerns, some countries explicitly cover non-scientific concerns in their regulatory process. Argentina, one of the first adopters of GM crops and now one of the largest producers, only authorises crops for cultivation if they are deemed to meet three criteria. There needs to be evidence that the GM crop doesn't have significantly different effects on the agro-ecosystem to its non-GM equivalent. The same goes for food safety – there must be evidence that food derived from the GM crop is as safe and nutritious as non-modified equivalents. Finally, there needs to be assurance that the crop won't encounter domestic or international commercial restrictions. Even if a GMO passes the safety assessment, approval can be denied if it could have other adverse consequences for the country.

In Asia, many regulatory systems are still in development. Although India's 1986 Environmental Protection Act provides the basis for the GMO regulatory framework, the Food Safety and Standard Authority of India is still in the process of formulating specific regulations for GM food. Currently applications for cultivation of new GM crops are assessed by a committee from the Ministry of Environment, Forest and Climate Change, but a bill has also been proposed to create an independent biotech regulatory authority.

Only in 2009 did China introduce its first comprehensive Food Safety Law, creating a Food Safety Commission to oversee food safety monitoring. Whilst the government has stated an ambition to speed up innovation and application of agricultural biotech, the commercial prospects don't match the investment in research. Currently the responsibility for GMO approval lies with the Ministry of Agriculture, which assesses safety for humans and the environment. Approval of a crop for cultivation and marketing is a lengthy process, with a minimum of four permits being required.

A single agency is responsible for food regulations in both Australia and New Zealand, assessing all applications for GM food on a case-by-case basis. Currently cotton and canola are the only crops grown, but a wide variety of food and feed are licenced for import.

In other areas, companies don't have the luxury of combined approvals. Biosafety laws often differ even between neighbouring countries, so producers have to incur the cost of going through the regulatory process in each of the countries. One of the international attempts to harmonise international GM regulations is the Cartagena Protocol on Biosafety, which focusses largely on ecological impacts. The agreement has been ratified by 170 parties, including the European Union and 166 UN member states, but excluding Australia, the USA and Canada.

Greg is helping signatories in the developing world to meet their obligations, as many of them don't have a system in place. Of the 47 countries in sub-Saharan Africa, just 18 have national biosafety laws. He's found that some countries aren't open to GM technology and take a very precautionary approach, so he focusses on countries which are keen to judge the technology on its merits. "I can help people set up regulatory systems so they can decide for themselves whether crops are safe," he said. "It makes sense to pick places where there's a political will and a scientific capacity to do that."

Another issue which he sees with international regulations is asynchronous approvals between countries. Greg said: "I think we have to figure out a way for the international community to reduce the time it takes for countries to make approvals. And countries may have to rely on approvals made by other countries when we're talking about small amounts of grain mixed in with other grains."

As things stand, different countries don't even necessarily have the same definition of GM – a crop which doesn't need to be approved in one country could still need to be approved in another. In particular, we are still waiting to find out how the EU chooses to class some of the new breeding techniques discussed in Chapter 4.

Another big question being asked around the world is whether it is right to regulate GM crops differently to conventional varieties. The safety of new varieties made using non-GM breeding techniques can be examined by national food authorities, but this isn't always the case. Generally, they go without the rigorous environmental and health assessments for GM organisms and foods.

This approach is being questioned in the EU, including in a recent report from the UK Council for Science and Technology, which advises the Prime Minister on scientific issues. The authors concluded that a special GM crop regulatory process would be justified if GM technology is 'especially dangerous' or is so poorly understood that there are 'extraordinary uncertainties'. The evidence, it concludes, doesn't point towards danger or extreme uncertainty. The European Academies Science Advisory Council (EASAC) has likewise called for a new regulatory system that focuses more on the nature of the end product rather than on the process by which it was made.

Penny Mapplestone, who we met in Chapter 13, is amongst the many Europeans who are frustrated with the situation they find themselves in. "I would love to see a more enabling framework in terms of technology," she said. "It's fantastic how molecular techniques have developed over the last few decades, and it's really disappointing that we can't take advantage of them in Europe."

The perversity of the current situation is most striking when you consider glyphosate-tolerant crops. Most are created using GM techniques, yet Cibus (the genome-editing company we met in Chapter 4) plans to bring out glyphosate-tolerant flax. The flax will come with exactly the same benefits and challenges as GM varieties, but without the legal requirement for regulation. The simple answer isn't to regulate genome-edited products in the same way as GMOs, because conventional breeding has also brought us herbicide-tolerant crops.

There is certainly a lot of scientific support for regulating new products in the same way regardless of how they were created. Although he wouldn't object to the system, Greg doesn't see a political will to do this. He said: "I tell people all the time that our regulatory systems are not solely about the science. They don't just ensure safety, they build consumer confidence in independent assessment."

As a student in the mid-1980s, Greg wrote a law paper about challenges facing GMO regulation, and it is striking that many still apply today. When he's helping to develop regulatory systems, one issue is who will be the primary regulator within that country. He explained the problem: "There are issues with the USDA, for example, being both a promoter of biotechnology and a regulator. Conflicts of interest can arise when you have the same agency doing both of those things."

In the decades since the first GM plant was created, the science has moved on in ways nobody could possibly have imagined. The politics, however, needs more time to catch up.

Chapter 16

The Right to Choose

As we shape the future of our society, including what role GM technologies may or may not play, we have to consider more than just the science. The views of individual members of society are a vital component of our choices, yet we have to deal with an extreme diversity of views. At the start of his Nobel Prize acceptance lecture, economist Amartya Sen asked the question: "Is reasonable social choice at all possible, especially since, as Horace noted a long time ago, there may be 'as many preferences as there are people'?"

In a democratic society, 'who decides' is a fascinating question, and there's a risk that the answer is sometimes 'whoever shouts the loudest'. It's pretty clear what the loudest voices are saying in the GM debate – what's harder to discover is the diversity of views held by the quieter majority.

The UK Food Standards Agency does regular Public Attitudes surveys on issues surrounding our food. When participants were asked about their food safety concerns in November 2015, the top three were food hygiene when eating out, food poisoning, and food hygiene at home. GM foods came in at 8th place, with 22% of respondents expressing concern. An interesting aspect of this survey was that participants were first given an open-ended question to state any concerns they thought of, before being given a list to choose from. If you only look at the initial spontaneous responses, the percentage of people expressing concern was much lower. In this case the list was topped by food hygiene when eating out, next came the use of additives, and this was followed by GM foods (with 5% of respondents expressing concern). It's not clear whether the participants who didn't rate GM foods as a concern did so because they saw them as low risk or because they felt the strict EU regulations were appropriate.

Eurobarometer surveys look at attitudes in the EU states, and in 2010 participants seemed optimistic about 'biotechnology and genetic engineering'. When asked about the likely effect of these technologies on the way of life in 20 years, over half of respondents expected them to have a positive effect, 11% expected no effect, and only 20% expected a negative effect. Attitudes were noticeably less positive when consumers were asked about the specific application of biotechnology in GM foods. In 2010, 23% thought that GM food should be supported, while 61% disagreed. This is slightly less positive than the 2005 survey.

Given that such a high proportion of US farmers have accepted GM you might expect US consumers to be more positive, but it seems that many are still sceptical. A 2014 Pew Research survey asked American participants some questions on the safety of GM foods, and found rather negative responses. A minority of adults (37%) say that eating GM foods is generally safe, while 57% say they believe it is unsafe and the rest responded 'don't know'. Interestingly, there was both a gender divide and also a divide between participants with different levels of education. Men were more likely than women to believe GM food to be safe, and people with higher educational qualifications were also more likely to be convinced of its safety. What's also worth noting is that there was an even more negative attitude towards pesticides – with 69% of people responding that eating foods grown using pesticides was unsafe.

So why have GM foods gained a negative perception? One reason is perhaps the political climate at the time they were introduced. GM foods arrived in the wake of BSE, one of the UK's largest food scare stories, which had undermined trust in both government officials and scientific experts. As Rothamsted Research's Professor Nigel Halford explained: "This enabled pressure groups to dominate the debate, and European consumers have been bombarded with inaccurate information, half-truths and wild scare-stories, as well as bizarre and frightening imagery. Even if consumers do not believe the more hysterical of these stories, why should they take the risk of buying GM food products?"

Studies of whether European consumers will genuinely avoid buying GM products have been very enlightening, and reveal the gap between what people say and what people do. In one study, identical fruits were labelled either 'organic', 'conventional' or 'spray-free GM' at roadside stalls in Sweden, France, Belgium, the UK and Germany. When the prices were equal, 'organic' fruits were the top sellers whilst 'spray-free GM' had around 20% of the market share. However, things abruptly changed when 'spray-free GM' fruits were sold at a 15% discount, and the 'organic' fruits were sold at a 15% premium. The 'spray-free GM' market share rose to 43% in Sweden, 33% in France, 30% in the UK and 36% in Germany. This doesn't match the researchers' findings from questionnaires – in Germany, for example, only 12% of respondents said they would buy the 'spray-free GM' fruit even at the discount. Just as with other food choices, it seems that price has a big hold on us.

This survey does, however, show that some people will avoid the GM label. So should GM foods be labelled in our shops? Given the level of distrust in GM foods, it isn't surprising that there is such a dedicated lobby to promote mandatory GM labels.

Celebrities have lent their support, with Gwyneth Paltrow backing the group Just Label It. In a speech delivered at Capitol Hill, Washington DC, she told her audience: "I'm not here as an expert. I'm here as a mother, an American mother, that honestly believes I have the right to know what's in the food I feed my family."

It sounds logical (well, not that an actress should influence FDA policy, but that she should have the right to know what she feeds to Apple and Moses). However, when we start to unpick what a 'right to know' would mean, things become a bit muddier. Does she have the right to know everything she wants about her food, or just information which is relevant to its safety or nutritional value?

She clearly doesn't feel she has the right to know every detail about what's in her food – nobody has ever asked for a sequenced genome plus the chemical structure of the enzymes and starch molecules found in a kidney bean. So, we're selective about what information we have the right to know, and what makes it onto the list? Do we have a 'right to know' something simply because we want to? If enough people wanted to know whether the workers who made their food were paid a fair wage, would that mean they had the right to a mandatory living wage label? I'd be up for that. But what if they wanted to know what colour the factory door is? After all, green doors are bad luck... And if we had this information, would it actually affect which products we chose to buy?

Connected to the 'right to know' argument is the 'right to choose'. As the situation in Europe has shown, labelling hasn't brought about choice. Instead, the effect of labelling has often been not to increase choice but to push GM foods off the shelves. In reality, pushing GM out of the market has generally been the intention of those campaigning for labels. In Europe it has allowed campaign groups to send 'gene detectives' in search of GM labels, then put pressure on retailers to remove these products.

It's not even clear how many people are interested in GM labelling. Polls in the USA show that an overwhelming majority of Americans will answer yes to the question of whether they want GM foods to be labelled. However, this has been criticised for a number of reasons – they generally aren't given an 'I don't know' option, and there's no indication of how important the issue is to them. Even if there's interest, that doesn't mean there's knowledge. There have been American surveys where a quarter of respondents said they'd never heard of GMOs, or where respondents showed overwhelming support for spoof policies such as 'label all food with DNA in'. You also get a very different result depending on how you design the poll. If you ask people what information they would like to see on food labels, only a few percent mention GMOs.

Even if GMO labels are mandatory, there is still the question of exactly what form that information would come in. Gwyneth has the right to know that FDA guidelines allow trace amounts of insect in chocolate, but individual chocolate bars don't have to come with a label to say whether or not they actually have bits of wing in them. Some people would like a GM label on every individual item, whereas others would like access to that information somehow if they chose to look. One possibility is a QR code you could scan for more information, although the fashion for QR codes has been extremely short lived.

Opinions aren't just divided between those who oppose GMOs and those who promote them, there's also a divide in the pro-biotech camp. There's a line of thinking which considers it misleading to label foods as GM. The GMO ingredient has only been approved for consumption if it is considered safe, yet a label could suggest otherwise. People in this camp would often like to see more valuable information prioritised on the label, such as the presence of allergens and nutritional information.

Others believe labelling is the transparency needed to regain public trust. Anyone already sceptical about biotechnology could react to the industry's arguments against labelling as proof they have something to hide, that they aren't interested in protecting consumers. Some proponents of this argument believe the outcome of the debate in Europe would have been different if consumers had been treated with more respect. The tomato paste which arrived on UK shelves as the first GM product was clearly labelled, and the UK intended to continue a policy of GM labelling. This didn't last long as the food industry soon faced an (apparently unexpected) dilemma. Shipments of soybean and maize arrived from the USA in late 1996 with small amounts of GM material mixed in, around 2%.

Nigel explained the dilemma: "Retailers had a choice: either label everything containing US soybean or maize as potentially containing GM material, or abandon the labelling policy. It chose the latter, a decision that, with hindsight, was undoubtedly a mistake because consumers felt that GM food was being introduced behind their backs."

Whether or not an initial labelling policy could have averted a consumer outcry, a belated move towards labelling doesn't seem to have abated it. The EU now has a labelling policy for GM foods, though there is so little on the market that this label is seldom seen. It's also worth noting that, despite the EU's strict labelling policy, a surprisingly large amount of the GM crops we grow don't end up in a product which requires labelling. For a start, meat or dairy products from animals which ate GM feed doesn't need to be labelled.

There is also a threshold in the EU of 0.9% GM material in non-GM foods to allow for small amounts of accidental mixing. This strict 0.9% for labelling has also been adopted in Australia, New Zealand, Saudi Arabia and Kazakhstan. In total, over 60 countries officially label GM, though sometimes this is only on foods containing over 5% of GM ingredients. Of course some countries which have the requirement for labelling in reality lack the regulations to enforce it.

In no country do you have to label animal products from livestock fed GM feed, although it is still possible to avoid these foods by buying organic. A hotly-debated exemption is GM crops which go through processing to make a product containing no DNA or protein. Processing means that glucose syrup, vegetable oils and sugars from GM crops are completely identical to products made from non-GM varieties. A label is therefore perverse if you're worried about GM on health grounds, though not if your concerns are ethical or environmental. Perhaps 'derived from' needs to be added to the GMO label for accuracy, although how many consumers will understand the distinction is another matter.

Another common exemption is food sold for immediate consumption. Kenya's labelling laws, for example, do not apply to foods sold by restaurants and food vendors. A big complexity when considering labels in Africa is the amount of food that is sold by market traders, much of which doesn't have a label at all.

One distinct drawback of labelling is the potential expense. Studies in America about the cost of GM labelling have come up with wildly different answers, partly because of the assumptions they are based on. We don't know, for example, when food manufacturers would choose to switch away from GM ingredients in order to avoid the label. This could lead to the extra cost of new product formulation and sourcing of new ingredients. If GM foods are going to remain on the shelf alongside non-GM foods, the supply chain will need to be overhauled. Transporting and storing grains separately will be an expensive endeavour, and this has implications even for African countries considering whether to grow GM crops for export. There would also be high costs of monitoring compliance (something which isn't necessarily done effectively in all countries which currently boast labelling laws).

Maryland farmer Jennifer Schmidt explained her take on how the complexity of the supply chain would lead to high costs of segregating GM and non-GM grains: "For people who haven't followed a corn or soybean seed from the field to the fork, it would seem to be an easy thing to just put on a label. For those of us in farming, we know that it's not so simple and would have a catastrophic impact on our family farms. There is nothing simple about the food supply chain, from a commodity grain grown in my field to a food ingredient used in a baked good, cereal or other item on your grocery shelf."

Difficult, however, doesn't mean impossible. We successfully segregate oilseed rape varieties intended for industrial uses from those used in food, for example. Still, Jennifer sees the expense as a major concern: "Segregation is costly. We know because we do it every year, year in and year out, and have for years. We do it because we get paid a premium for ensuring that the specialty grains and seeds we grow are 'identity preserved', very much as with the certified organic process. This requires us to use some of our grain tanks for segregation. It requires us to do more 'house-keeping' – cleaning equipment trucks, trailers, planters, harvesters, grain bins, etc. – all along the food supply chain to ensure that we have preserved the identity of that crop. It is an inherently more costly system."

Given the potential costs of labelling, it's hardly surprising that corporates involved in the food business have something to say about it. Corporates which have lent support (vocal and financial) to the anti-labelling lobby include Coca-Cola, PepsiCo, Nestlé and the major seed companies. But even within the world of large corporations there is disagreement. In 2016, Campbell Soup announced plans to label all its products, and has voiced its support for a mandatory labelling bill in the US.

For some, labelling is an opportunity, in-keeping with brand values such as naturalness or transparency. In most cases, the promoters of labelling are planning to go GM free. The list of brands which have supported labelling include those whose parent companies fund the anti-labelling lobby; in striking contrast to its parent company Unilever, Ben and Jerry's ice cream is a member of the Just Label It campaign. Perhaps most perverse is the announcement that PepsiCo will label Tropicana orange juice non-GMO, even though there are no GMO oranges on the market. Although hardened Orwellians might substitute the term doublethink, fans of 1984 will no doubt appreciate the accusation from Just Label It's CEO, Gary Hirshberg, that the company is engaging in "Orwellian corporate doublespeak".

Those companies who choose to label their foods will likely transfer any increased cost to the consumer, and you could argue that those who support labelling should be the ones to pay. In reality, nobody is particularly keen to know that a food contains GM ingredients – those interested in GM labels are actually looking for non-GM food. The option of a reliable non-GMO labelling scheme seems like a way to ensure those people who want to eat GM-free foods can access them, while those who aren't interested don't have to pay the price. Whole Foods is one of the multi-nationals to see a premium market in GMO-free labelling, and its commitment to GMO transparency by 2018 has triggered many of its suppliers to reformulate their products.

Sellers are only likely to adopt GMO-free labels if they believe the expense of a segregated supply chain will be offset by increased sales or charging a premium. Monopolising on this will mean a more prominent label to attract consumers, and shouting louder about being GM-free could give an even more misleading impression than a mandatory policy for GM labelling. The case of Tropicana is just one example of a label being not simply misleading but intentionally deceitful. That isn't the only GM-free label on a product with no GM equivalent – one Texas-based company actually labels its salt as GM free. Even when the label does allow consumers to distinguish between GM and GM-free foods, they might be encouraged to pay more for their food because the label intentionally gives the mistaken impression of a safer product. Perhaps the label needs to come with information about GM safety studies, whether this is a disclaimer or a QR code of its own?

If brands are going to adopt labelling policies, they need to be backed up by reliable information. The Canadian government, for example, has developed a national standard for voluntary labelling, although it allows foods to be labelled GM free if they contain up to 5% GM ingredients by weight. This surprisingly high percentage is the same in the US, a fact I suspect no company's commitment to transparency will stretch to highlighting.

A key question is does a GM label give valuable information? It depends what we mean by valuable. Someone wishing to avoid GM foods would say it is very valuable, it allows them to choose alternatives. This doesn't make the information meaningful though. For a start, you need the knowledge to understand the issues surrounding GM. And those people who do have more understanding of GM will potentially need more information before they find a label useful. Virus-resistant and herbicide-resistant crops come with different issues, so to be really valuable information would have to include details such as type of modification. What would be wonderful to see are multiple measures of sustainability, not necessarily related to GM, and social measures such as the farmers' share of the profits. If GMO labels come with huge practical challenges, more detailed labels hardly bear thinking about, particularly if you're sourcing from smaller farms. It does, however, reveal the inadequacy of a simple GMO label.

If you are marketing a 'natural' or 'wholesome' product, then you may want to go beyond GM-free labels to actively promoting mandatory labelling. Anything which casts doubt over the safety of alternatives can only be a good thing. Interestingly, key players in the pro-label lobby include groups such as Citizens for Health, whose mission is to promote 'natural health choices – choices other than therapeutic drugs, surgery and radiation'. In this case it isn't clear which businesses fund its GM campaign, although we do know that its anti-corn-syrup campaigns were funded by The Sugar Association.

Even if a GM-free image can be good for business, companies may be going GM-free because their executives believe there are benefits for the consumer or society. Ben and Jerry's Social Mission Activism Manager Chris Miller said: "We didn't make this decision based on marketing. Given our commitment to family, small-scale agriculture, it felt like the right decision." Although he added: "Clearly the fact that this has become a trend helped push us in that direction."

In terms of the price implication, he said: "We did not raise the price of a pint of ice cream in transitioning to non-GMO because it did not impact the cost of ingredients for us in a significant way. I can tell you we didn't take a margin hit by changing our sourcing."

Just because Ben and Jerry's didn't put their prices up it doesn't mean the same will happen with other products – Ben and Jerry's ice cream is a premium product in which the main ingredient is milk from cows fed on GM grain.

Overall, the future of GM labels seems as unclear as the future of GMOs themselves. In 2015, the US congress passed a bill to prevent states from creating their own GM labelling laws, but the delayed decision on whether to label AquAdvantage salmon shows it is still a hot topic. We've seen the benefits and dangers of other labelling schemes, and GM labels risk being misused in a similar way. Labels advertising foods as 97% fat free, for example, entice consumers with a healthy product, while information on the sugar content is shared in small text on the back. With this kind of tactic often employed for labelling, it is not in any way clear how we could use GM labels to empower people to make choices rather than manipulate them into paying more.

The labelling debate has raised relevant questions about the importance of respecting people who hold different points of view, even if you believe the views themselves fly in the face of the evidence. Many scientists are acutely aware of a tendency to alienate people by failing to listen to other points of view, something many people in the GM debate have been guilty of. Even the use of the word 'public' can be telling. It immediately gives an impression of 'them and us'. In fact, we are all members of the public, and everyone who eats has a stake in how their food is grown. It risks suggesting a fundamental difference between scientists and others, and ignoring the huge diversity in people who might make up 'the public'.

Some have argued that even using the words *opinion* or *perception* suggest a dismissal of people's views. Bart Gremmen, Professor of Ethics in Life Sciences at Wageningen University in the Netherlands, describes why he believes perception is the wrong word: "I think it is often used to contrast scientific objectivity, saying 'OK, scientists and technologists, we are objective, we have knowledge, and there is the public with its perception'. When they use the word they immediately allude that it's just a perception." Opinion, he points out, can be interpreted as 'just an opinion amongst others'.

The fault, it seems, is not just amongst scientists. Vicki Hird, an environmental consultant and former senior campaigner at Friends of the Earth, told me: "GM campaigns in the past may have been panicky in tone but they were based on the precautionary principle and talking to the public in terms they can understand. They were making sure we are all protected against tech that was not proven to be safe in every sense – direct and indirect. That was reasonable against a backdrop of inadequate data."

If these campaigns genuinely had been using terms which allowed more people to understand the data that existed and the data which campaigners believed to be lacking, I would have applauded them. As it stands, I don't think that the powerful images of mutant Frankenfoods left society with an understanding of the evidence. Patronising consumers with 'terms they can understand' doesn't suggest a dialogue based on information, instead it speaks of passing your own views in one direction.

This raises the interesting question of whether it's an environmentalist's responsibility to decide that there is inadequate data and present the scare story, or to provide information on data that exists and data that is lacking. Environmental NGOs have been hugely influential in the GMO debate, and as a result have attracted criticism.

As with many complex issues, few people spend much time looking at the evidence. Decisions are instead based on what feels right and on the views of the people you trust. Eroding trust in your opponent has become a very popular pastime in the GM debate, with 'scientific experts are paid by Monsanto' being a favourite line. Conspiracy theories are rife, and campaigners on both sides like to tell you that their opponents are pulling the wool over your eyes. For environmental groups, this tactic is a good one. If you're going to be duped, most people would choose to be duped by over-zealous environmentalists rather than by boardroom executives cackling while you lap up the lies. Unfortunately, that means that some campaigns of misinformation have become very successful, and people have been duped.

It would be completely wrong, however, to disregard the views of the environmental movement as a whole. Alongside a certain amount of damage, environmental groups do some fantastic work around the world, which includes spreading knowledge in developing countries. With GMOs, however, the trouble is often the quality of the knowledge.

In 2015, British charity ActionAid criticised its Ugandan subsidiary for saying that GM technology potentially causes cancer. ActionAid Uganda had been showing farmers pictures of Séralini's rats with tumours (as discussed in Chapter 4), and had commissioned radio adverts warning of the dangers of eating GM foods. Many concerns have been raised about the way anti-GM campaigns have looked to shape attitudes in the developing world, although how we can empower people in the developing world with accurate information is an unanswered question. As we'll see in the following chapter, lack of information is a common theme in the story of GM in Africa and beyond.

Rebecca Nesbit

Chapter 17

GM in the Developing World

"Africa is in danger of becoming the dumping ground for the struggling GM industry and the laboratory for frustrated scientists," is the warning from GRAIN, an NGO supporting small farmers. Yet others disagree. A report by Information Technology and Innovation Foundation, an American Think Tank, accompanied its figures about increased income for small-holders adopting GM crops with the conclusion that: "Anti-GMO activists have erected barriers to agricultural biotech innovation that could cost the poorest nations up to $1.5 trillion through to 2050."

If you look at the reasons behind GRAIN's warning, you may start to become slightly sceptical. Many of the arguments simply focus on herbicide-tolerant crops, and they state fears that GM will usher in terminator technology. However, it's also true that even if activists retracted their opposition, it still wouldn't be plain sailing to guaranteed rewards. How developing countries could realise the benefits of GM crops while minimising the risks is a complex question; we can't simply introduce seed varieties from the west and expect these to benefit the world's poorest farmers.

Already, millions of farmers in developing countries are planting GM seed. In 2015, GM crops were widely grown in Asia and South America, and in Africa they were planted in Burkina Faso, Sudan and South Africa. The impacts have, of course, been variable. The scale of the current losses from pests and disease mean there is great potential for GM crops, but the ways they

261

have been introduced mean the results haven't always been positive.

Developing countries in different parts of the world face very different situations. Since 1960, per capita production of cereals has grown hugely in Asia and South America, yet it has actually declined in Africa. Asia and South America have benefitted from modern seeds while Africa still relies largely on traditional varieties. It's hardly surprising that there's little commercial motivation to develop technologies for Africa: the USA's seed market is about 1,200 times the size of the market in Uganda. This gap could be filled by philanthropic or government-funded research, and there are some interesting projects in the pipeline. So far, however, we've sometimes seen the introduction of inappropriate varieties which have been developed for other countries.

Glenn Davis Stone, Professor of Sociocultural Anthropology & Environmental Studies at Washington University, has studied the effect of GM crops on small farms in the developing world. His first experience of GM technologies was right at the beginning of commercial production: "I remember taking a break from my work to walk down Broadway. I went into a greengrocers and I saw something called a Flavr Savr tomato being sold. It was advertised as a genetically modified tomato, and I didn't have a clue what to make of that." A taste test back in his apartment revealed little difference between the Flavr Savr and a conventional tomato.

He didn't think much more about it, until consumer backlash started to grow. The advertising campaign launched by the biotech industry in the late 1990s reignited Glenn's interest, as conflicting claims were made about whether these crops would feed people in the developing world or endanger them. He explained: "I couldn't help but notice that the overwhelming majority of what was being written was coming either from companies that had a vested interest in it or from activists who also had a vested interest in it. And basically there were a lot of claims being made about how these technologies would affect farmers and the consuming public, and there was almost no role in the debates for people who actually studied those things."

Small farms came up repeatedly, so as the 20th century drew to a close he mothballed the research he was doing and shifted his focus to GM crops. Although he sees clear benefits for big commodity farmers, he's more sceptical about claims that it will be a key technology for feeding the third world.

As far as Bt cotton is concerned, Glenn reports that it has been an added tool to fight pests, but the overall impact in the field has been pretty variable: "The activists' claims it's been a big failure are not very credible – in many cases it's been a success. The biotech companies' claims that it's been just a smashing, remarkable success are also overstated."

This is in the context of a troubled cotton industry. Hybrid cotton seed is common in India, so farmers were buying new seed each year even before the arrival of GM. Poor regulation of the seed industry means the seeds can be unreliable, and many are mislabelled. Deceptive labels have sometimes given Bt a bad name, with farmers buying varieties such as 'BEST cotton' and finding it isn't actually resistant to insects.

The same is true elsewhere. In Uganda, for example, trust in commercial seeds has been undermined by counterfeiting of modern crop varieties combined with poor quality controls for new seeds. This is confounded by a quickly changing seed market, with a confusing new array of seeds on the market. The result is that farmers aren't equipped with the information they need to choose the best seed for their land.

Bt cotton has added a new element to the confusing mix, especially with multiple Bt genes now available which offer resistance to different pests. Glenn said: "What I'm concerned with is that, at the cost of some added insect protection, farmers have had to deal with even more rapid change in the technology and even more unrecognisable technologies."

There's also the inevitable issue of pests developing resistance. Continuously introducing new genes which pests aren't resistant to may work in the developed world, but will it work for Indian farmers? If they can't access local seed varieties with a working resistance gene in, they will be without the protection they had come to rely on. Relying on genetic resistance could mean farmers lose the skills needed to control pests in other ways. Skill, Glenn passionately believes, is an essential part of sustainable farming.

One thing that Glenn has noticed is that some of the claims made by anti-GM activists can play into the hands of the biotech industry. In the region of India he works, one of the more outlandish claims is that sheep are being killed by grazing on the stubble of Bt cotton. With unsubstantiated claims such as this, it's easy to diminish the credibility of legitimate concerns by fighting back with 'these activists are always coming up with crazy ideas'.

Across continents in Africa, Dr Klara Fischer was finding some parallels. Klara, from the Swedish University of Agricultural Sciences, spent time living in a South African village to study the role of agriculture in the wider environment. Here she spoke to farmers about Bt maize and the government policies which had made it available. It was a study she approached with an open mind: as an undergraduate in the 1990s she had received very conflicting views from her lecturers. Geneticists had focused on the possibilities, whilst ecologists had highlighted the risks. In reality, she found both.

When she first spoke to the farmers it immediately became clear that they weren't interested in talking about Bt maize – it was just another maize variety to them. Instead, they wanted to discuss 'the Programme', the state-run project which had introduced Bt crops. This was expensive for the government, which was providing small-scale farmers with highly-subsidised seeds. What the government didn't provide was knowledge. Klara explained: "Theoretically Bt could be of value in the region because stem borer caterpillars are a problem, and people can't afford pesticides. But information was so limited that the farmers didn't even know that the crops were resistant to stem borers."

The farmers received marketing which covered both Bt maize and herbicide-tolerant maize, and they didn't understand the difference between them. Even when farmers planted insect-resistant maize, they mistakenly thought someone would come and spray their crops with herbicide.

Another problem was again a common one: the crop varieties which the Bt genes had been added into weren't right for small-holder farmers in the local conditions. Klara said: "The varieties were suitable for large-scale industrial farming, where everything is optimised – you have enough fertiliser and enough water. That's not the conditions this maize was planted in."

Klara tried to understand why this situation had occurred, and her conclusions were related to attitudes she had encountered. She said: "People think small-holders are inferior and they don't know how to do things. There was also the idea that development is linear and should be the same for everyone. In this context, development was for the small-holders to become exactly like the large-scale farmers."

These attitudes meant a focus on how to change the farmers, without looking at the circumstances they were farming in. There was a belief that farmers needed to become more business-minded, although the reasons small-holders weren't seeing their farms as a business couldn't be solved with seed. Klara explained: "They don't have any storage, and they don't have a tarmacked road to the market – they don't have anything to make it possible for agriculture to be a business."

Ultimately these problems are nothing to do with Bt technology. If Bt genes had been introduced into appropriate maize varieties, and if the programme had come with education, the benefits might have been realised.

As things stand, Klara is concerned that the focus on seed technology distracts us from social and political issues. Her research shows that farmers struggle with many other, more serious constraints than having access to the latest seed varieties. She therefore doesn't see GM crops as a key solution to the problems facing African farmers. "This is not to say that GM crops should be kept away from smallholders, though if they are to be useful for poor farmers, they need to be allowed to save and recycle seed," she said. "But that means we can't control where the seeds are spread – it's a tricky thing to solve."

For each crop we need to understand what the risks would be of widespread release, and put these in the context of potential benefits. In the case of Bt crops, the main danger would be insect-resistance. Elsewhere, this risk is managed by planting non-GM crops as a refuge, yet this kind of management practice is impossible to enforce if seed is saved and freely traded. The mix of crops often grown by small-holders reduces the risk of resistance, but we still have to think about the effect on farmers if resistance became widespread.

This balance between regulation and freedom is an important issue for new types of GM crops being developed with Africa specifically in mind. These projects are designed to benefit small-holders rather than to produce tightly-regulated seed to sell for a profit. As a result, projects often take place at public sector institutions, with the involvement of multi-national businesses limited to the donation of intellectual property. Much of the funding is philanthropic, from organisations such as the Gates Foundation, USAID and the UK's Department for International Development.

The Gates Foundation is a major funder, and has developed the concept of Global Access to help ensure their projects make a difference to the lives of people in need. To receive Gates Foundation funding, grant holders need to sign a Global Access Agreement. Whilst the scientists (or at least their institutes) are able to profit from their work in the developed world, any products and information must be made widely available at an affordable price to the people the Foundation is working to help. The Agreement also stresses the importance of projects being promptly and broadly disseminated, to help people as fast as possible.

There's a very long list of GM crops being developed in and for Africa, at an expense of many millions of dollars. We discussed some of these in Chapters 7 and 9, and there are far more out there. Insect-resistant cowpea, salt-tolerant rice and virus-resistant bananas are some of the many crops undergoing trials, and some of them have been for years. Political pressure and poor regulatory systems means these crops often stall in the approvals process, becoming stuck at the field trial stage without being released. An analysis by the UK's Royal Institute of International Affairs (more popularly known as Chatham House) concluded that: "This 'convenient deadlock' of continual field trials allows governments to manage political risks by effectively balancing the demands of pro-GM and anti-GM lobbies – proponents of GM have a pipeline of technologies, while opponents are appeased by the failure of any to gain approval."

Of the projects in the pipeline, some of the most interesting involve cassava. The fact it can be vegetatively propagated eliminates some of the worries which accompany the seed industry, and the number of people who rely on it means improved varieties could have wide benefits. Cassava is now a dietary staple for hundreds of millions of people, and is grown mainly by small-holders. It was brought to Africa alongside the slave trade, and Nigeria has become the world's largest cassava producer.

As we saw in Chapter 9, cassava plays a huge role in food security, yet it has seen a fraction of the research investment of the largest commercial crops. Cassava brown streak disease, caused by two viruses, is devastating crops in East Africa, and GM provides some possible ways to tackle it. The disease was first detected in 1936, at which point control programmes relegated it to very small pockets of disease. This changed in 2000 when it began to spread rapidly in Eastern Africa, leading to crop loss on a huge scale.

Dr Laura Boykin, a computational biologist at the University of Western Australia, explained the severity of the problem: "A lot of farmers are facing a really long hunger season in East Africa. We have this perfect storm, in that we have this crop that feeds so many people and so many people rely on it, and these viruses are just devastating the yields. The situation in East Africa is as important as Ebola has been in West Africa."

One tactic to reduce these losses is using biotechnology to develop virus-resistant cassava varieties. The technology is very similar to Hawaii's virus-resistant papayas, which we discussed in Chapter 7, and field trials are underway with partners in Uganda and Kenya. Once they have been through the regulatory approval, scientists intend to distribute the GM cassava royalty-free to farmers and local breeders. The breeders will be free to introduce the resistance into all the varieties which farmers want.

Another possibility is to tackle the spread of the disease, which in this case means tackling the insect vectors. The disease is transmitted by whiteflies, tiny insects which suck sap from the underside of leaves. Much of Laura's work has focussed on whitefly biology, and this is feeding into the work of scientists looking to genetically-engineer cassava to produce whitefly resistance. Laura is also working with conventional breeders, although sadly natural insect resistance from other cassava varieties isn't transferable to east Africa. Resistant varieties exist in South America, but these turn out to be resistant to a different whitefly species.

Creating whitefly-resistant cassava is a long-term project, while farmers are losing their crops now. To tackle the problem immediately, Laura and her colleagues are educating farmers so they understand that whitefly transmit the disease. What the education programmes certainly don't give is advice to control the whitefly with pesticides. Laura said: "Chemical controls and whitefly are a deadly combination because they become insecticide resistant very quickly. Even if farmers could afford chemicals, which they can't, they wouldn't be useful."

Laura, who has been named one of 12 badass women scientists by the TED Fellows programme, finds working with African scientists to be a thrill. "The best thing I've ever done in my career was just going 'I'm now going to be working on this with these people, because it matters'. And when you've made that decision, everything about your day is better. I don't pretend to know what smallholder farmers need, but I know my collaborators do, and I listen to them."

This recognition of the value of local knowledge is essential if the developing world is going to benefit from biotechnology. Not only will the products be more likely to meet people's needs, they are more likely to be accepted if they come from local providers. Collaborations with African researchers are becoming more common, as are programmes which fund African scientists to study at the world's top scientific institutions. Still, lack of scientific capacity was an issue identified by researchers at the University of Toronto when they did interviews in sub-Saharan Africa.

Interviewees came from organisations such as regulatory institutions, research centres, farmer organisations, the media and NGOs. Many of them identified lack of training and expertise as a barrier. Without the local skills for development of technologies, it won't be possible to eliminate the concern from some interviewees that foreign biotechnologies were "... some effort by the Western world to come and take advantage of poor Africans."

Elsewhere, there is the equivalent concern that GM opponents from developed countries are inappropriately exporting their views to Africa. Professor Giles Oldroyd, who is working on the nitrogen-fixing crops we discussed in Chapter 7, is concerned that the farmers who could benefit the most from genetic modification are being denied this: "Personally, I think it's a real tragedy, and it's a reflection of organisations like Greenpeace campaigning very aggressively, and really promoting untruths in the developing world. They're creating a state of fear which to me is very colonialist."

These concerns are shared by Professor Diran Makinde, Director of the African Biosafety Network of Expertise, based in Ouagadougou, Burkina Faso. He said: "The activists meet policy makers one-on-one, and pump them with all sorts of misinformation. That is what we're grappling with in Africa."

To make GM more acceptable in Africa, Diran sees it as important to have local funding sources as well as local expertise. He speaks passionately about the need for governments to invest more in science and technology, and he laments that many influential Africans prefer to sponsor football clubs. A shortage of funding has left little room for African innovation, meaning there is still a lack of local ownership. Diran does see a new opening: "One thing that is good is that we are getting off-patent genes which our local seed companies can take advantage of."

There still needs to be strict monitoring to ensure the safety and quality of local biotech seeds, but as patents expire African breeders are free to use technology they may not be able to create themselves. Whereas modifications such as the addition of Bt genes have been mostly limited to commercial varieties, once the genes are off patent it becomes much easier to introduce them into local crop varieties.

Of the many GM crops being developed for Africa, some are tackling problems we have failed to tackle by other means. However, we shouldn't forget the solutions which already exist. We invest vast quantities of money into new technologies, yet existing knowledge and technology often isn't being used to its full potential. If knowledge was spread and advice was followed, many farmers around the world could improve their productivity.

In parts of sub-Saharan Africa, 80% of farmers are women, and lack of access to education can be a persistent barrier for them. Women are less likely than men to own land or livestock, adopt new technologies, use financial services, or receive an education or advice from university extension services. Beyond agriculture, women's education has many social and economic benefits. Particularly relevant to discussions of food security, women's education is associated with lower birth rates.

Research by the UN's Food and Agriculture Organization concluded that closing the gender gap in agriculture would generate major gains for the sector and for society. If women had the same access to productive resources as men, they could increase yields on their farms by an estimated 20 or 30%.

Access to resources needs to come alongside access to education – there are plenty of examples of high-tech solutions failing because they haven't come with the appropriate training. Dr Paul Seward, Director of Farm Input Promotions Africa Ltd. (FIPS-Africa), explained the situation his team faces in East Africa: "A typical farmer does not produce enough food, and her family goes hungry for up to six months per year. She cultivates by hand small plots of land ranging in size from 0.1 to 2 hectares."

FIPS-Africa aims to make appropriate farm inputs such as fertiliser more accessible to small-scale farmers. In Kenya, they set up a system of village advisors. Initially, the donors – the United States Agency for International Development, the Rockefeller Foundation and the UK Department for International Development – wanted Paul's team to promote disease-tolerant varieties of maize from emerging seed companies, along with better fertiliser.

Two years later, the team was surprised to find that adoption rates were low, unlike the local varieties of sweet potato. Paul said: "Farmers told us that maize was difficult to grow: it required expensive seed and fertiliser, and was sensitive to drought and the striga weed."

Now the village advisors offer farmers a whole range of improved crop varieties which meet local needs, including bananas, vegetables and fruit trees. Still, Paul and his team didn't give up on the new varieties of maize. Instead they are helping farmers do small experiments of seed and fertiliser to see whether the maize will be appropriate for them. He said: "In 1996, I discovered by chance that resource-poor farmers in Siaya County, in Kenya, who had never before used fertiliser, wanted to purchase it in small 100-gram packs costing only US\$ 0.1. Thousands of farmers purchased the small packs, experienced the benefits on their farms in an affordable way, and then asked for larger quantities (1–10 kilos) to improve food security."

Simply supplying small packs of seed and fertiliser doesn't work on its own – farmers need advice if they are going to make effective use of modern inputs. Many farmers in Kenya were placing two to five seeds in a hole and, if they used fertiliser, placed it directly on top of the seed. This leads to very low productivity, so Paul and his team developed a 'planting string' to help village advisors teach farmers better planting. It's simply a piece of string with four bottle tops clamped on it to measure how far apart single seeds should be, and a small card to measure the distance between the fertiliser and the seed. "It can be rigged up in a couple of minutes and costs a few cents," said Paul. "This simple and inexpensive tool has helped thousands of farmers to increase their maize crop productivity up to fivefold through better seed spacing and fertiliser placement."

For GM critics questioning the need for high-tech solutions, this kind of initiative is an obvious choice for their support.

Overall, stories of GM in the developing world speak of both problems and promise. The take-home message seems, once again, to be the importance of looking at crops and situations on a case-by-case basis. Even more than in the developed world, we need to consider the whole context, and to be aware of the social situations.

Norman Borlaug, father of the green revolution and recipient of the 1970 Nobel Peace Prize, has been famously quoted as saying: "I now say that the world has the technology – either available or well advanced in the research pipeline – to feed on a sustainable basis a population of 10 billion people. The more pertinent question today is whether farmers and ranchers will be permitted to use this new technology? While the affluent nations can certainly afford to adopt ultra low-risk positions, and pay more for food produced by the so-called 'organic' methods, the one billion chronically undernourished people of the low income, food-deficit nations cannot."

It is certainly true that anyone without enough to eat can ill-afford to be idealistic. We just need to be careful not to simply shed the 'natural is best ideology' for 'technology will fix it'. The evidence of what will bring benefits in a specific context will need to take into account environmental and social conditions. It's complex, but the evidence is what we're going to need.

Chapter 18

Do We Need to Grow More Food?

For thousands of years, humans have been changing crops and farming practices so they can produce more food. Even now, how to increase yield is a question that farmers, scientists and breeders around the world constantly ask themselves. Yet, on a global scale, will this actually help us feed more people? There are two conflicting facts flying around about whether we need to increase food production at all. Firstly, we hear that the world actually produces enough food; it just doesn't reach the right people. In complete contrast, the UN predicts that farmers need to produce 70% more food by 2050. Some estimates are even higher – researchers at the University of Minnesota, for example, have more recently put the figure at a 100–110% increase. So who's right, and does it matter?

Before we answer that question, it's worth noting that 'we don't need any more food' isn't a complete argument against GM, although it is often used that way. Many of the GM crops we considered in previous chapters have different aims, such as enhanced nutrition, reduced inputs or protection against yield loss in years of drought. Still, whether we need to increase yields is a fundamental question in debates over which GM technologies might be useful. In fact, given the environmental impacts of growing food, it is a vital question to be asking ourselves anyway.

The simple answer to 'do we produce enough food to feed the world's population?' is 'yes': in 2015 we produced globally about 2,900 calories per person per day. That is, however, only the simple answer and is for today's population level. The levels are very different if we look at individual countries.

At the country level, this calorie count is based on food available for human use after taking out all non-food uses, such as exports, animal feed and seed for the following year. It is extremely variable between countries, with Turkey, Argentina, Egypt and Cuba having over 3,800 calories available per person while Zambia and the Central African Republic have less than 2,000.

The UN figure of a 70% increase goes beyond this current day estimate and takes into account future needs. This is a vital addition to our calculation, though on closer inspection the exact figure seems to be rather suspect. Nonetheless it has been widely used in FAO documents and the media, with some variations. In a 2011 report it was presented as: "Toward 2050, rising population and incomes are expected to call for 70% more food production globally, and up to 100% more in developing countries, relative to 2009 levels."

Fearing that the 70% figure had assumed a life of its own, the authors of a 2012 FAO report felt the need to put the record straight. For a start, they revised the estimate down to a 60% increase needed from 2005/2007 to 2050, based on a more accurate figure for what production had been in 2005/2007. They also cautioned that figure itself isn't actually very meaningful – it is based on volume and compares very different products (such as oranges, grain and meat). On top of this, there are lots of uncertainties in the prediction, and the figure is based on what would be needed to match the projected demand as we think it may develop. We don't know how many people there will be alive in 2050 or what they will choose to eat. Their food choices will be important, particularly in regards to animal products. It is much more efficient to eat grains than to feed them to animals for meat, yet as people get richer their diets tend to include more meat and dairy products.

I asked Tim Wheeler, Professor of Crop Science at the University of Reading, whether this figure stands up to scrutiny: "I would say that it is a good analysis of food trends, backed-up by the best agricultural datasets in the world. However, as with all projections forward in time, the figure of 70% is underpinned by assumptions based on past trends of how the world may change in the future. I would take it as a good first approximation that there will be an agricultural production challenge in the future, whilst recognising that the actual figure may not be exactly 70%."

FAO reports based on this figure argue that we can achieve this increase in food production given enough investment and appropriate policies for agricultural production.

So it seems that it's a pretty safe bet we'll need more food in the future, but we also need to understand why our current production levels are still leaving people hungry. According to the FAO, high food availability is associated with relatively low levels of undernourishment in most regions. However, high food availability doesn't always mean high food security. As economist Amartya Sen noted in his 1998 Nobel Prize speech: "Famines can occur without any major decline – possibly without any decline at all – of food production or supply."

In the most dramatic example of the politics of food, extreme famines are a feature of dictatorships. Elected governments are under too much pressure to allow a famine to develop, and can implement measures such as emergency public employment, whereas both communist and capitalist dictators around the world have allowed their countries to descend into famine.

While much of the literature on famine investigations focussed on food production, Sen brought to the foreground the problem of purchasing power. Even when food is plentiful, people will go hungry if they can't afford to buy it. He has warned against seeing hunger as simply a food problem, and instead advocates understanding it as a wider economic problem.

Of course food supply and purchasing power are linked; food shortages can lead to rising food prices which leave many people unable to pay. Sen isn't calling for us to replace one narrow outlook for another. Instead we need to broaden our focus beyond food production and consider the wider issues which cause almost 800 million people to go hungry. He said: "To prevent persistent under-nutrition, attention has to be paid to health care, in general, and in particular to the prevention of endemic diseases that deter the absorption of nutrients."

If everyone is to have a share of the food we produce, we have some distribution challenges. Food isn't divided equally between regions, within communities or even within families. At an international level, re-distribution will always be necessary – Singapore, for example, has no hope of producing all its own food. Where there is not enough local production to meet demand, trade has been vital in filling gaps. At the same time, trade comes with risks for local farmers and economies.

Former United Nations Special Rapporteur on the Right to Food, Professor Olivier De Schutter, believes that our crazy situation of nutrient deficiency and hunger existing alongside obesity has at its roots our emphasis on producing cheap calories. Over his two terms of office at the UN, the problem he found himself obsessing over was the cycle of hunger in the Global South. Small-scale farmers are displaced by competition for resources, while government support focusses on agriculture for exports. As city slums swell with the influx of displaced farmers, governments become increasingly reliant on importing cheap food. This cheap food makes it even harder for farmers to make a living.

Farmers in the global South can find that markets are flooded by cheap food from overseas, making it difficult for them to have a livelihood. Exporting food from developing countries is also a problem, as Olivier explained: "The other complaint is that we capture the resources from the South to serve our own needs and this is made possible thanks to the vastly superior purchasing power of the populations in the North. The two phenomena converge to encourage in the South export-led agriculture to satisfy our own needs, but at the expense of the ability for small-scale farmers to make a decent living."

Education, healthcare, trade policies, and access to markets are just a few of the challenges intertwined with malnutrition. Tackling these challenges will be essential to ensure that the food we have available reduces hunger in the developing world. Meanwhile, we should perhaps also look at how to decrease demand in the developed world.

One reason that it is hard to predict the increase in production is that future uses of the crops we grow will change. It's also a reason that the figure predicts demand growing faster than the population. We don't just eat our agricultural produce, we wear it, drink it, smoke it, burn it for energy, make car tyres with it... even use it to get high.

A major destination for food crops is livestock, which is a much more resource intensive way to feed ourselves than simply eating the grain ourselves. Olivier has major concerns about our meat consumption: "Industrial livestock production is unsustainable from the environmental point of view, and is diverting scarce resources, in terms of cereals produced, to feeding livestock in ways that are completely inefficient. Just picture that by 2050, 50% of cereals produced in the world will go to feeding animals."

Not all meat is equally damaging to the environment. Beef, for example, is worse than chicken, and grain-fed beef tends to be worse than grass-fed beef. Indeed, there are some situations where livestock is an integral part of a food system. Some Japanese farmers, for example, use ducks to control insects in their rice paddies. Still, we can't avoid the fact that it takes far more calories to raise an animal than we get from eating it. As a result, Olivier believes that it will be essential to eat less meat not just different meat. He said: "In the EU, the average meat consumption is about 95kg per year per person. It's about 120kg in the US, Australia and New Zealand, it is 60kg in China. We should move towards a 30 or 33kg per year per person as a global average for meat production to be sustainable."

Around the world we see economic growth leading to an increased consumption of meat and dairy products. This is a major reason why production is projected to increase, as more people choose diets similar to those in the developed world. This increased demand is clearly going to be an environmental problem, yet it would be a moral problem if we didn't want other nations to enjoy the same diet as ours.

Olivier sees inequality between nations to be an issue: "Overconsumption of meat in rich countries is only possible because we have the purchasing power in Europe, in the USA and increasingly elsewhere that allows these populations to use massive areas of land and water in the global South to feed their increasing demand for meat. If we had a more equitable distribution of resources we would not have the luxury demands of the richest competing against the basic needs of the poorest."

The topic of the most environmentally-friendly way to provide people with enough protein is a hot and complex debate. It isn't just meat that we would need to consume less of, it's other animal products too. Cows consume grain and produce methane (a potent greenhouse gas) whether they will be used for milk or beef. There are some more radical alternatives to the meat and dairy we currently rely on.

One of my favourites is entomophagy, the practice of eating insects. Scientifically it's a no brainer. Insects have the environmental benefit that they need less food to produce the same amount of protein. They also have the health benefit that they are generally lower in fat than traditional meats. The trouble is that most people in the developed world recoil at the thought, even though they are unwittingly consuming insects in products such as chocolate and orange juice. My personal experiences of eating crickets, moth pupae and beetles have been variable, mostly depending on what they are cooked in. Elsewhere, these delicacies are an important part of diets – the FAO estimates that over 2 billion people regularly eat insects. The question is whether entomophagy in the west can become more than a quirky fad and contribute to our global food security.

Similarly, we have already discussed lab grown meat, and there are also more realistic alternative sources of protein including products made using plants proteins such as tofu, or ideas like Quorn. Made in fermentation tanks using a fungus, Quorn has a lower carbon footprint than meat and needs fewer inputs. The fungus still needs to be fed with grain, but is much more efficient at converting the energy into protein than animals are. Products made using soya protein themselves are perhaps most relevant; feeding humans soya-based products, rather than growing animals using soya grain, is clearly much more efficient.

A further, more obvious solution to reducing the need for food production is to stop throwing so much of it away. Globally we create about 1.3 billion tonnes of food waste each year. In the UK alone, we throw away 15 million tonnes of food annually – the average UK household throws away the equivalent of six meals each week. In the USA, about 1,249 calories worth of food is lost per person per day.

It is well within the power of consumers and producers to drastically reduce food waste. We need to create innovative food packaging, not be so fussy about the appearance of our fruit, stop 'buy one get one free' offers which encourage shoppers to buy too much, and understand the difference between 'use by' and 'best before' dates.

The level of waste is shocking, and given the environmental cost of producing food it seems criminal that we throw so much of it away. It does, however, illustrate why 'we don't need to grow more food' isn't an argument against all uses of GM. From the very first GM food, the Flavr Savr tomato, to one of the most recent, the Arctic apple, there are products out there explicitly designed to ensure products last longer to help us reduce food waste.

Even worse than food that is thrown away is the food we waste by consuming it when we don't need it. Obesity has become an epidemic, causing strain on health services and preventing many people from living full and active lives. At the turn of the century the number of overweight people caught up with the number of people who are underweight. This followed a slow decline in the number of people who are hungry alongside a rapid increase in obesity. In the much of the Western world, society is set up to encourage poor health choices. Workplaces are often sedentary, our supermarkets stock cheap sugary foods, and adverts use psychologically-developed techniques to encourage us to buy them.

None of these factors are pre-decided – we have the power to change the demand for food. Should we wish, we also have the power to change the size of the human population, whether this is by empowering women in developing countries to control their own fertility or thinking about how many children we have ourselves. The issue is summed up by the Optimum Population Trust, a charity promoting both those paths: "On a finite planet, nothing physical can grow indefinitely. The more of us there are, the fewer resources there are for each of us and for members of other species with which we share the planet." At the moment, the people alive today are living in a way which will create challenges for those not yet born.

Chapter 19

The Latest Innovation in a Broken System?

Is modern agriculture an incredible achievement which under-pins an otherwise unimaginable civilisation, or an environmental catastrophe bringing about a mass extinction of species? Well, in many ways it is both. It supports a huge population, mostly in a state of health which our pre-agricultural ancestors could only dream of. It has freed people from work in the fields, allowing us to dedicate resources to everything from medicine to art. However, not everyone is receiving a fair share of this bounty, and the environmental impacts will be felt by generations to come. It has turned forests into fields and is a major contributor to greenhouse gas emissions; it is degrading natural resources, and polluting air, soil and water.

It's no exaggeration to say that we are facing an extinction crisis, and that agriculture has a lot to answer for. Wildlife is suffering severe declines around the world, and some of the data showing this come from my back garden. Each night, moths are attracted to the bright bulb of my moth trap. This is part of the Rothamsted Insect Survey, and data from the network of light-traps led Butterfly Conservation to conclude that: "Across Britain, the total abundance of larger moths declined significantly, during the 40-year period from 1968 to 2007."

We don't fully understand the causes of decline, but it seems that agricultural intensification is a major culprit. We see this pattern repeated across different groups of wildlife, and the

International Union for Conservation of Nature (IUCN) identifies agriculture as the main threat to 87% of globally-threatened bird species. Quite simply, with one quarter of the Earth's terrestrial surface now under cultivation there's less space for wild plants and animals.

So far I have been comparing GM crops to the current state of agriculture. But is this all we aspire to? It's clear that our current system, whilst amazing in the number of people it feeds, doesn't feed the global population in a sustainable way. What's not in any way clear is how to achieve this. Solutions are likely to vary around the world, and there is no single answer. We need to consider the social and economic aspects of our food system, and consider both what needs to change in our methods of production and what needs to change in the nature of our consumption.

There are various reasons why 'business as usual' is a risky approach for the future of agriculture. So far, population growth hasn't led to widespread hunger. Agriculture has adapted, and we have continued to increase yields. Hunger is falling even as the population is increasing. Yet we can't take this for granted as the population continues to climb. We're already seeing the impact of climate change on farming, particularly shifting rainfall patterns. The drier regions of Africa are especially vulnerable to the potential for increased drought. It's also clear that continuing to increase yields in a 'business as usual' fashion will be bad news for many of the species we share our planet with.

One solution put forward for a more sustainable farming system is agroecology which, at the most basic level, means applying ecological science to agriculture. It seeks to mimic natural processes, and considers the entire agricultural system, not just the plant. Agroecology practices look to reduce reliance on external inputs, both for environmental and economic reasons.

Practices can include diversifying the crops we plant, and recycling nutrients and organic matter.

There are some impressive examples of situations where applying practices of agroecology has led to an increase in yield. Malawi, for example, has had success using nitrogen-fixing trees, particularly the apple-ring acacia. Also known as the balanzan tree or winter thorn, the tree has the unusual characteristic of shedding its leaves in the rainy season. This means that, rather than shading the crops, the nitrogen-rich leaves are dropped to fertilise the soil. A training programme has already supported 120,000 local farmers planting trees with their crops. Maize yields more than doubled as a result of the programme, and the trees themselves can be used for food and fuel. The flowers are even popular with bees.

This kind of approach is particularly important in areas where poor infrastructure and high prices mean farmers don't have reliable access to fertilisers. For the poorest farmers, a cheaper way of providing nitrogen to their fields can increase their stability and self-sufficiency. However, it isn't an either/or situation. Yields increased even more if the agroforestry approach was combined with the application of limited amounts of mineral fertiliser. This is a perfect example of science being used to discover ways to increase yield whilst decreasing inputs.

Former United Nations Special Rapporteur on the Right to Food, Professor Olivier De Schutter, has been keen to stress the importance of knowledge: "Many people dismiss agroecology as a dream of the romantics, as a return to the past, a nostalgia for the agriculture of our grandparents. No, it's agriculture for the future. I very much hope, and I believe, it shall be the kind of knowledge intensive agriculture that our children will be performing. Agroecology is demanding because it requires training, it requires knowledge. It requires knowledge to be spread from farmer to farmer in horizontal ways, rather than techniques being imposed top down."

Agroecology can clearly achieve great things in many situations, increasing yields whilst reducing inputs. However, there should be some words of caution. If reducing agricultural inputs leads to a reduction in yields, this could actually be environmentally damaging. For a start, it is important to consider the inputs needed to produce a certain amount of food rather than to farm a certain area. We also have to be careful about how much land we use for agriculture.

Whilst many conservationists have studied ways to support farmland wildlife, Andrew Balmford decided to investigate whether this is the right approach. A Professor of Conservation Science at the University of Cambridge, Andrew believes that looking at yield is fundamental when considering the least environmentally damaging way to produce our food. There are two broad schools of thought about how to reduce the impact of farming on wildlife: land sharing and land sparing. In a land sharing scenario, we should look at ways to make agricultural land friendlier to wildlife, even if this means lower yields. This outlook has received lots of political support recently, and the EU was spending around £5 billion per year paying farmers to take part in agri-environment schemes. In contrast, proponents of land sparing suggest that the better option is to maximise yields in the area you farm so that there is more land spare for nature.

To tackle the sharing vs sparing question, Andrew and his colleagues have done studies looking at trees, birds and insects, and in locations such as Poland, Ghana and India. They have found a remarkably consistent pattern between countries: more species benefit from land sparing than from land sharing. Many species struggle to survive even in low-intensive agriculture, so the best option is to maximise the amount of natural habitat.

Interestingly, this pattern becomes even more pronounced if you want to produce more food from an area, as we very likely will in the future. Even at current production levels, the studies found that land sparing is still the better approach. Andrew said: "Even if, by tackling food waste, population growth and diet, demand were somehow to fall, land sparing would still be the least bad option."

Andrew's work comes to a different conclusion to some other studies. In these alternative studies, the scientists have looked at the number of species in different habitats, and found that medium-intensity farmland can still support plenty of species. Andrew explained why this is: "Many sensitive, narrowly-distributed, forest-dependent species have been replaced by weedy, agricultural-tolerant, cosmopolitan species which are common elsewhere."

If you simply count the number of species on a small area of farmland, you ignore the fact that the same species are being found on farmland around the country. Meanwhile, many other species are losing their habitat to farms. Generally, the species of greatest conservation concern are those which need very specific habitats, and often have a narrow global distribution. These specialist species have proved to be the most likely to do better under highly-intensive farming and land sparing. The purple emperor butterfly, for example, spends its time in woodland canopies. However well farmland is managed, it won't be a suitable habitat for the purple emperor or other species with such specific requirements.

Land sparing has the potential to bring other environmental benefits too. Andrew has also collaborated on work suggesting that spared land could offset some greenhouse gas emissions from agriculture.

Andrew is keen to stress that focussing on high yields doesn't mean simply continuing with our current trajectory of food production. It is still important to research ways of increasing yield without increasing environmental impact. He sees any way of increasing yields as worth considering, and this includes genetic modification. "I believe as conservationists we can't afford to be ideological about what to consider," he said. "We have to be agnostic, and compare options based on what counts."

For Andrew, what counts is largely yield, but also the negative impacts of the high-yield farming. Ultimately, what counts is the overall environmental impact of producing the food we need.

Andrew is confident that the evidence shows that land sparing is the least bad option in theory, and now he would like to see us trying it out in practice: "While we've spent very large amounts of money trying to get land sharing to work, we've spent more-or-less nothing on implementing land sparing."

The problem with implementing land sparing isn't necessarily so much in increasing yields as in ensuring this actually leads to land being spared. Market forces alone are unlikely to do this effectively, and we risk simply increasing production if farming becomes more profitable. Although in the past yield increases have sometimes led to spared land, this isn't always the case. There are, however, some ingenious examples of schemes which have simultaneously increased yield and spared land. For example, villagers around Bandipur National Park in India used to offset crop losses to elephants by illegally grazing their cattle in the Park. A scheme which helps farmers fence their fields has increased yield by keeping elephants out, and has simultaneously meant that entering the Park isn't the most valuable use of the farmers' time.

In many situations, however, higher yields won't automatically lead to land sparing. Governments and consumers may both have a role to play, and possible ways to ensure land is spared include certification schemes, protected areas, subsidies and taxes. Now that the potential benefits of land sparing have been understood, it seems like the time to develop policies which would ensure these benefits are realised.

Scientific research and new agricultural practices are essential if we are going to improve our food system, but they are not the full story. If we have a more transformative agenda, we need to consider markets as well as farming. On one level, agroecology has become quite mainstream. It can be a way to adapt the current food system – it's even mentioned on Syngenta's website. However, other people have a more radical interpretation. Agroecology can also include a social movement, tied in to the concept of food sovereignty. It can be associated with socioeconomic and political principals which questions the basis of our entire agricultural system.

The idea of food sovereignty was launched at the World Food Summit in 1996. It goes beyond food security because, after all, you can be food secure even if you're in prison. The concept was developed by the organisation La Via Campesina, an international peasant's movement which now represents about 200 million farmers and agricultural workers around the world. It works to support small-scale agriculture, and "strongly opposes corporate driven agriculture and transnational companies that are destroying people and nature".

Food sovereignty is both everyone's right to sustainable, healthy and culturally-appropriate food, and people's right to define their own food and agricultural systems. It ensures that the rights to land, water, seeds, livestock and biodiversity are in the hands of producers and not the corporate sector. Promoting food sovereignty has become La Via Campesina's main goal.

Central to food sovereignty is the theme of partnership between producers and consumers. It seeks to bring farmers and consumers closer together, and shorten the food chain from farm to plate. It focusses on rural development and strengthening communities, promoting initiatives such as farmers' markets, farm stores and cooperatives.

Another aspect is participatory research, where farmers have a greater involvement in shaping research agendas. Scientists don't simply transfer knowledge to farmers; the two groups collaborate to share knowledge. There are also calls to have a greater involvement of farmers and consumers in deciding how governments allocate their research funding. Agro-economist Dr Gaëtan Vanloqueren, from the University of Louvain in Belgium, is passionate about the social potential of a greater focus on the views of farmers and consumers. His outlook is that "agroecological research for transformation is part of a bottom-up, participatory process in which farmers and citizens take centre stage."

Food sovereignty questions the free market. In free-market economies, prices are set through supply and demand. Prices are simply related to what people are willing to pay, without interference from governments or other authorities. Instead, the food sovereignty movement aspires for people to be free to shape their food systems through, for example, tariffs and subsidies.

It also challenges capitalism by favouring cooperatives and social enterprises rather than corporations. In a capitalist corporation, control is with the people who bring in the capital. It's a method of governance which Gaëtan is critical of: "Those who bring their workforce have no say in the firm's governance, including strategic decisions about the use of the yearly economic surplus. Should it be allocated to investors or to workers, or reinvested in the organisation for its development? There are alternative types of firms, such as cooperatives, governed in a democratic way – one person, one vote instead of one dollar, one vote."

Some of the principles of agroecology have been incorporated into organic farming, although it doesn't always meet the standards of the food sovereignty movement. Currently, organic farmers will often rely on inputs such as biopesticides, which may be sold by multinational companies. Also, some organic growers, distributors and sellers are themselves part of larger corporations. A look at the parent companies of organic brands reveals some corporations with very different images. It's a long list, but to name just a few: Green and Black's chocolate is owned by Mondelez International (formerly Kraft Foods), Seeds of Change is owned by Mars, and Honest Tea by Coca-Cola.

An interesting issue to consider alongside agroecology and food sovereignty is employment. Some agroecology techniques are labour-intensive rather than highly mechanised, and this has an implication for farm jobs. Mechanisation can free people from hard manual labour, creating time for education and other employment. However, in parts of the world where unemployment is high, agriculture can create much-needed jobs.

Agroecology is not necessarily a farming system which is entirely reliant on a large human workforce, although it is sometimes viewed that way. Gaëtan said: "Agroecological approaches are perfectly compatible with a gradual and adequate mechanisation of farming. However, one tractor replacing the daily work of twenty landless labourers is only progress if nineteen jobs are created in the secondary and tertiary sectors."

There are many social and geographic factors which affect the creation of jobs, and 'machines taking our jobs' remains a widely debated topic. If we take the developed world as an example, mechanisation of agriculture has often coincided with an increase in jobs. This was the striking finding of an analysis of almost 150 years of UK census data by economists at Deloitte. The UK's employment figures have increased in this time, leading to the conclusion that technological change has coincided with the creation, not destruction, of work. This isn't just a historic trend. The loss of employment in agriculture and manufacturing since 1992 has been more than offset by rapid growth in the caring, creative, technology and business services sectors. Overall, the authors conclude that when a machine replaces a human this results in faster growth and, ultimately, rising employment.

In the long run, if food sovereignty includes the ability for citizens to choose how their food is produced, it is up to societies to decide on whether they want to increase or decrease the need for employment in agriculture. In Chapter 17 we discussed Klara Fischer's work with South African small-holders. For many of the people she spoke to, farming was a back-up. What they really wanted was a job. Elsewhere in the world this isn't always the case – some people may see agriculture as their passion, yet have to leave farming because they can't make ends meet.

How GM crops could fit into an alternative food system depends on your perspective and the crop you're talking about. Some GM crops have been associated with the problems which agroecology is hoping to tackle, such as monocultures. Glyphosate-tolerant crops are definitely not compatible with food sovereignty, both because of their reliance on the input of herbicide and their origin with multi-national businesses. On the other hand, technologies for nutrient enhancement, provided they are royalty free, aren't necessarily incompatible with agroecology or food sovereignty.

Many agroecology advocates don't see GM foods as compatible with their ideology. Agroecologist Professor Stephen Gliessman is certainly very critical. He speaks out against industrial agriculture, which relies on over-use of resources such as water, and says of biotech crops: "The extraordinarily rapid and extensive adoption of GM crops since 1996 is a clear indicator of how perfectly genomics fits in to the industrial model. It takes the strategy embodied in the development and marketing of hybrid seeds and moves it one step further."

However, the same isn't necessarily true of future GM crops. Some GM crops are designed to tackle the same problems as agroecology is, such as the need for large amounts of water and chemical pesticides. As a result, there are supporters of the need for both genetic engineering and agroecology. Two very vocal advocates are Professor Pamela Ronald from the University of California, Berkeley, and her husband Raoul Adamchak, an organic farmer.

Pamela explained their position: "We believe that the judicious incorporation of two important strands of agriculture – agricultural biotechnology and agroecology – really are key to help feed the growing population in an ecologically balanced manner. We feel very strongly that pitting genetic engineering and organic farming against each other only prevents the transformative changes that we need on our farms."

Technology is agnostic – a lab technique can't be inherently in conflict with a sustainable system. The conflict arises in part because of its cost, which can favour large, rich corporations.

After the 2012 Rothamsted protest I spoke to two of the activists, students who had travelled on the protest bus. It was an insightful chat and we were left with a much greater understanding of each other's opinions and motivations. It surprised me that they felt they were protesting against capitalism, and GM seemed simply to be a scapegoat. Our conversation made me wonder whether the GM debate is just diverting energy away from the real objective of reforming the whole food system. Banning GM crops would do nothing to change a system of capitalism.

I couldn't help but see an analogy in nature. During my PhD I spent time studying butterflies in Gibraltar, and my field site was next to a bird of prey rehabilitation centre. Each spring, a spectacular stream of buzzards, eagles, vultures and the occasional stork would cross the sea from North Africa as they headed to their summer breeding grounds. Fearing for their chicks, angry yellow-legged gulls would mob the vultures; occasionally our friends at the rehabilitation centre would rush out in a speed boat to rescue a vulture which had been forced down into the sea. Actually the vultures were just scavengers migrating through and posed no threat to the gull chicks. So by extending their energy mobbing the vultures, the gulls were simply diverting their resources away from the real task in hand: rearing their young. In the same way, the real debate about capitalism and corporate control got forgotten amid the police horses, chanting and organic pizza.

The genuine arguments apply to our whole food system, whether or not GM is involved, and they raise some valuable questions. What do we aspire to in our food system? Will we achieve most by working to completely reform the system, or by making improvements to the system we have? Even if we believe a system should be fundamentally overhauled, it doesn't mean we shouldn't look to improve it in the meantime.

It is interesting to consider how we can change the system to better protect the environment and benefit society as a whole. There are advocates of changing regulations and government policies, affecting farmers and businesses along the food supply chain. There are many government subsidies which have attracted criticism from some parties, particularly those which reduce agricultural diversity by encouraging farmers to grow a particular kind of crop. Wise subsidies, however, have the potential to promote responsible farming. There's also potential for consumer-led, bottom up approaches.

In many cases, the knowledge required for more sustainable farming practices exists. Unfortunately, providing this knowledge for farmers doesn't always make a good business, and so state funding can be needed for education. State funding of agricultural research is also essential, something which has often been squeezed by funding cuts and hampered by short-term grants. Industry funding of agricultural research has been welcome to fill gaps left by a shortage of government funding, yet it focusses on research which fits into a model of agriculture led by big business.

Our current system is both destructive and remarkable, and its positive and negative effects are unequally distributed. If we are going to keep as many of these benefits as possible whilst tackling the problems, we need to be open to different ideas. There will be a need for compromises, and respect for other points of view. We face decisions which need to be informed by evidence from science, economics and social science. Whether or not we want our changes to be radical, if we're looking to tweak our system or overhaul it, it's clear that we can't let the market decide. Everyone has a say in how we shape the future of our food system; it affects not just our own lives but those of generations to come.

Chapter 20

Conclusion

Love it or hate it, Twitter provides a fascinating social commentary. The hashtags I follow include #GMO and #GMcrops, partly for the information they lead me to and partly because I am amused by the difference between them. If you search for #GMO you may well learn that Monsanto is attacking democracy, whereas #GMcrops is more likely to bring you news of farmer benefits.

Humans are well known to suffer from confirmation bias, especially when considering emotionally-charged issues. We seek out information which confirms our existing outlook, we interpret ambiguous information to fit our views, and we even selectively remember the evidence which is most convenient for us. What better tool than Twitter to help us find news stories, reports and experts to support our existing views about GM foods?

The GM debate is a clash of ideologies, with complex evidence to challenge everyone's world view. The difficulty lies in accessing the most reliable evidence, and making sense of any conflicts. When I looked at the scientific literature, there was information which re-enforced my existing views and information which challenged them. Despite the complexity, it has brought me firmly to one conclusion: every crop and every situation is different, and we need to assess them on a case-by-case basis. There are many benefits and risks to consider, and many people who are set to gain or lose. Ultimately, nothing is all good or all bad (in fact, that's the conclusion I've come to in life!).

I've also become increasingly aware that the accusations against GMOs often apply to non-GM crops as well. Genetic modification can be intimately connected with many of the challenges facing our food supply – loss of biodiversity, hunger, nutrient deficiency, climate change and dominance of large companies. Depending on how it's used, GM could address or intensify these problems. Two decades after GM crops were commercialised, isn't it time to focus on these huge challenges, and not chase GM as a scapegoat? It's definitely time to put sustainability over profit, but we need evidence to work out what sustainability really looks like and how we can achieve it.

Rob Wallbridge, an organic farmer from Western Quebec, speaks persuasively about the need to acknowledge the complexities of agriculture and to understand the similarities between different farming systems: "If we want to stop the food fight between pro- and anti-GMO factions, if we want more courtesy and civility in our food discussions, if we want to start addressing some of the most pressing concerns in food and farming, we must stop polarising the debate, begin to recognise the true breadth and nature of the gap, and respect the choices available, as well as the people who make them."

Addressing social and environmental challenges will be crucial if we are to feed a growing population in a sustainable manner, and we need resources to do this. UN research indicates that investment in agriculture is five times more effective at reducing poverty and hunger than investment in any other sector. Governments and philanthropic funders which support agricultural development all have to decide whether GM research and promotion is the best use of their money, and this decision needs to take both scientific and social factors into account.

There are legitimate arguments that some funds directed towards biotechnology could be better spent on alternative solutions. However, it's also worth considering what could have been achieved if funds used for anti-GMO campaigns had been used to support food security. Campaign groups can't be in control of where multi-national companies or the Gates Foundation invest their money – what they can control is where they put their own funds.

There are lots of options for anyone looking to put their energies into feeding more people with less damage to the environment. These include research into less energy-intensive farming, providing knowledge and resources to poorer farmers, more equitable distribution of food, reducing waste, or even transforming patent law. As individuals, we can also make changes to our lifestyles. Do we head to the supermarket to buy grain-fed beef shipped in from the other side of the world, to then chuck half of it in the bin, or do we make more sustainable choices?

None of these avenues are mutually exclusive with GM crops, but some could reduce the need for technological solutions. Weed ecologist Professor David Mortensen made an important point in Chapter 5: knowledge doesn't sell like products. What if we spread knowledge about good farming practices, rather than misinformation about GM foods? The UN's World Food Programme gives a sobering prediction of what could be achieved if we made better use of current knowledge and technologies: we could reduce the number of hungry people in the world by up to 150 million simply by giving women farmers the same access to resources as men.

There are clearly many dimensions of the quest to feed a growing population in a changing climate, and new technologies will no doubt be part of this. Biotechnology, in one form or another, is here to stay. The question we should really be asking is: how can we access the benefits while minimising the problems?

Index